LARS.
BOSWELL
KANOLD
STIFF

Algebra 1

Applications • Equations • Graphs

Chapter 8
Resource Book

The Resource Book contains the wide variety
of blackline masters available for Chapter 8.
The blacklines are organized by lesson. Included
are support materials for the teacher as well as
practice, activities, applications, and assessment
resources.

McDougal Littell
A HOUGHTON MIFFLIN COMPANY
Evanston, Illinois • Boston • Dallas

Contributing Authors

The authors wish to thank the following individuals for their contributions to the Chapter 8 Resource Book.

Rita Browning

Linda E. Byrom

José Castro

Rebecca S. Glus

Christine A. Hoover

Carolyn Huzinec

Karen Ostaffe

Jessica Pflueger

Barbara L. Power

James G. Rutkowski

Michelle Strager

ISBN: 0-618-02046-2

3456789-CKI- 04 03 02 01

Contents

Contents

Contents

Descriptions of Resources

This Chapter Resource Book is organized by lessons within the chapter in order to make your planning easier. The following materials are provided:

Tips for New Teachers These teaching notes provide both new and experienced teachers with useful teaching tips for each lesson, including tips about common errors and inclusion.

Parent Guide for Student Success This guide helps parents contribute to student success by providing an overview of the chapter along with questions and activities for parents and students to work on together.

Prerequisite Skills Review Worked out examples are provided to review the prerequisite skills highlighted on the Study Guide page at the beginning of the chapter. Additional practice is included with each worked-out example.

Strategies for Reading Mathematics The first page teaches reading strategies to be applied to the current chapter and to later chapters. The second page is a visual glossary of key vocabulary.

Lesson Plans and Lesson Plans for Block Scheduling This planning template helps teachers select the materials they will use to teach each lesson from among the variety of materials available for the lesson. The block-scheduling version provides additional information about pacing.

Warm-Up Exercises and Daily Homework Quiz The warm-ups cover prerequisite skills that help prepare students for a given lesson. The quiz assesses students on the content of the previous lesson. (Transparencies also available)

Activity Support Masters These blackline masters make it easier for students to record their work on selected activities in the Student Edition.

Alternative Lesson Openers An engaging alternative for starting each lesson is provided from among these four types: *Application, Activity, Graphing Calculator,* or *Visual Approach.* (Color transparencies also available)

Graphing Calculator Activities with Keystrokes Keystrokes for four models of calculators are provided for each Technology Activity in the Student Edition, along with alternative Graphing Calculator Activities to begin selected lessons.

Practice A, B, and C These exercises offer additional practice for the material in each lesson, including application problems. There are three levels of practice for each lesson: A (basic), B (average), and C (advanced).

Contents

Reteaching with Practice These two pages provide additional instruction, worked-out examples, and practice exercises covering the key concepts and vocabulary in each lesson.

Quick Catch-Up for Absent Students This handy form makes it easy for teachers to let students who have been absent know what to do for homework and which activities or examples were covered in class.

Cooperative Learning Activities These enrichment activities apply the math taught in the lesson in an interesting way that lends itself to group work.

Interdisciplinary Applications/Real-Life Applications Students apply the mathematics covered in each lesson to solve an interesting interdisciplinary or real-life problem.

Math and History Applications This worksheet expands upon the Math and History feature in the Student Edition.

Challenge: Skills and Applications Teachers can use these exercises to enrich or extend each lesson.

Quizzes The quizzes can be used to assess student progress on two or three lessons.

Chapter Review Games and Activities This worksheet offers fun practice at the end of the chapter and provides an alternative way to review the chapter content in preparation for the Chapter Test.

Chapter Tests A, B, and C These are tests that cover the most important skills taught in the chapter. There are three levels of test: A (basic), B (average), and C (advanced).

SAT/ACT Chapter Test This test also covers the most important skills taught in the chapter, but questions are in multiple-choice and quantitative-comparison format. (See *Alternative Assessment* for multi-step problems.)

Alternative Assessment with Rubrics and Math Journal A journal exercise has students write about the mathematics in the chapter. A multi-step problem has students apply a variety of skills from the chapter and explain their reasoning. Solutions and a 4-point rubric are included.

Project with Rubric The project allows students to delve more deeply into a problem that applies the mathematics of the chapter. Teacher's notes and a 4-point rubric are included.

Cumulative Review These practice pages help students maintain skills from the current chapter and preceding chapters.

Tips for New Teachers

For use with Chapter 8

LESSON 8.1

TEACHING TIP Teach students how to use their calculators to evaluate powers. Be aware that different calculators might require different keys to do this. Once students know how to evaluate powers, they can verify the properties of exponents by evaluating both sides of the equation. For instance, they can check that $5^3 \cdot 5^4 = 5^7$ because both sides equal 78,125. Remind students that they still need to know the properties so that they can simplify variable expressions.

COMMON ERROR Students might try to simplify the product of two powers that have different bases, doing things such as $x^2 \cdot y^3 = xy^5$. You can show them that they are wrong by plugging in values for the variables and evaluating both sides of the equation: they are not equal. Remind your students that they can only use the product of powers property when the two powers have the same base.

COMMON ERROR Ask students to evaluate an expression in which two powers are added, such as $3^2 + 3^5$. Some students will claim that the answer is 3^7. Others might say that this problem cannot be done because they have only learned properties of multiplication. Show students how to evaluate these expressions by evaluating the powers and then adding them. Then students can evaluate both $3^2 + 3^5$ and 3^7 to see that they are not equal.

LESSON 8.2

TEACHING TIP Zero and negative exponents can also be taught to students as an extension of the division properties of exponents covered in Lesson 8.3. You may wish to cover the division properties of exponents on page 463 first. Then it follows that

$$\frac{2^3}{2^3} = 2^0,$$

by subtracting the exponents. Since they also know that $2^3 = 8$, they could rewrite the problem as $\frac{8}{8} = 1$. Therefore, 2^0 must be equal to 1. A similar process can be used for negative exponents. If students know the division properties of exponents, then

$$\frac{x^2}{x^5} = x^{-3}.$$

The problem can also be written as

$$\frac{\cancel{x} \cdot \cancel{x}}{x \cdot x \cdot x \cdot \cancel{x} \cdot \cancel{x}} = \frac{1}{x^3}. \text{ Therefore, } x^{-3} = \frac{1}{x^3}.$$

COMMON ERROR Remind students that expressions with positive exponents are usually considered to be simpler than expressions with negative exponents. Therefore, students should always write their final answers with positive exponents. Review how to transform the negative exponents into positive ones to stop students from making mistakes such as $x^{-3} = -x^3$ or $y^{-4} = -4y$.

COMMON ERROR When students evaluate exponential functions they might make the mistake of multiplying first a, the initial value, and b, the growth factor, and then taking the power. Thus, for $y = 3 \cdot (2)^x$, when $x = 4$ they would incorrectly find $y = 6^4$. Remind students that the order of operations states that powers must be evaluated before multiplication.

LESSON 8.3

COMMON ERROR When using the quotient of powers property, some students always subtract the smallest exponent from the largest one. For instance, they might say that

$$\frac{3^4}{3^9} = 3^5.$$

Remind them that they must subtract the exponent in the denominator from the one in the numerator, even if this means obtaining a negative exponent for an answer.

LESSON 8.4

TEACHING TIP After multiplying or dividing expressions in scientific notation, students might have difficulty expressing their answer correctly in scientific notation. For example, they might know that

$$(1.4 \times 10^4) \cdot (7.6 \times 10^3) = 10.64 \times 10^7,$$

but they could have a hard time writing that number in scientific notation. Teach them a *"compensation law"* of scientific notation. This law states that if the number in front of the power of 10 must become smaller, then you must *compensate* by making the power of 10 larger. Since 10.64 > 10, you must make it smaller by moving the decimal point left one space. You must then *compensate* by making the power of ten one unit larger, 10^8. In a similar manner, this law can be used when the number in front of the power of 10 must become larger to be between 1 and 10.

INCLUSION Make sure to go over the new words included in this lesson such as million, billion, and trillion. You might want to write each of these in both decimal form and scientific notation.

LESSON 8.5

COMMON ERROR When evaluating exponential growth functions, remind students to write the growth rate in decimal form, even if it was given to them as a percent. Also, make sure that students remember to put the 1 in the function: some of them might incorrectly write $y = C(r)^t$ as their function.

COMMON ERROR When the growth factor is very small, students might graph exponential growth models as lines, because their graph closely resembles one. Remind students that the graphs for this type of function are always smooth curves, not lines. What students are seeing is the local behavior of the curve, which can be approximated by a straight line. You can demonstrate this fact with a graphing calculator by graphing any curve—a parabola, for example—and zooming in to show your students that if you get close enough to the curve, its graph always resembles a line.

LESSON 8.6

TEACHING TIP To introduce the lesson, talk about how the value of a computer falls over time. You can find real data for the depreciation rate for computers on the Internet. Use it to create an example where students have to find the remaining value of a certain computer in a given number of years after its purchase.

INCLUSION You might need to spend some time explaining the meaning of words such as appreciation, depreciation, inflation, and deflation, even with students whose first language is English. Make sure to use examples that students can relate to and understand.

TEACHING TIP Ask your students to make two separate lists, one for objects that appreciate in value and one for things that depreciate. Ask them to share their answers and discuss whether they are valid.

COMMON ERROR Students sometimes graph an exponential decay model crossing or touching the *x*-axis. Make a table with values large enough to show that even though *y* continues to decrease as *x* increases, the value of *y* never reaches zero. This means that the graph should never cross or touch the *x*-axis. You might even want to explain to your students what an asymptote is. Then they can see from their graphs and those in the book that the *x*-axis is always an asymptote for exponential models.

Outside Resources

BOOKS/PERIODICALS
Jones, Graham A. "Mathematical Modeling in a Feast of Rabbits." *Mathematics Teacher* (December 1993); pp. 770–773.

ACTIVITIES/MANIPULATIVES
Kincaid, Charlene, Guy Mauldin, and Deanna Mauldin. "The Marble Sifter: A Half-Life Simulation." *Mathematics Teacher* (December 1993); pp. 748–759.

SOFTWARE
Masalski, William J. "Topic: Compound Interest." *How to Use the Spreadsheet as a Tool in the Secondary Mathematics Classroom.* NCTM, 1990; pp. 16–19.

VIDEOS
World Population Review. Southern Illinois University at Carbondale, 1990.

NAME _____ DATE _____

Parent Guide for Student Success

For use with Chapter 8

Chapter Overview One way that you can help your student succeed in Chapter 8 is by discussing the lesson goals in the chart below. When a lesson is completed, ask your student to interpret the lesson goals for you and to explain how the mathematics of the lesson relates to one of the key applications listed in the chart.

Lesson Title	Lesson Goals	Key Applications
8.1: Multiplication Properties of Exponents	Use properties of exponents to multiply exponential expressions. Use powers to model real-life problems.	• Irrigation • Windmills • Multiple Choice Tests
8.2: Zero and Negative Exponents	Evaluate powers that have zero and negative exponents. Graph exponential functions.	• Basketball • Savings Accounts • Shipwrecks
8.3: Division Properties of Exponents	Use the division properties of exponents to evaluate powers, simplify expressions, and find a probability.	• Stock Exchange • Atlantic Cod • Learning Spanish
8.4: Scientific Notation	Use scientific notation to represent numbers, to perform operations with numbers, and to describe real-life situations	• Amazon River • Astronomy • Louisiana Purchase
8.5: Exponential Growth Functions	Write, use, and graph models for exponential growth.	• Compound Interest • Bicycle Racing • Bird Eggs
8.6: Exponential Decay Functions	Write, use, and graph models for exponential decay.	• Purchasing Power • Depreciation • Medications

Study Strategy

Planning Your Time is the study strategy featured in Chapter 8 (see page 448). You may wish to have your student post a calendar showing school assignments, extracurricular activities, and family commitments in a prominent place at home. This can be consulted for planning daily study time as well as for choosing when to schedule upcoming events.

NAME _____ DATE _____

Parent Guide for Student Success

For use with Chapter 8

Key Ideas Your student can demonstrate understanding of key concepts by working through the following exercises with you.

Lesson	Exercise
8.1	In order to test the business sense of her heirs, a rich woman made the following offer. "I will either give you a million dollars today or I will give you a dollar today, two dollars tomorrow, four dollars the next day, and so on for twenty days." Which was the better offer? Explain.
8.2	Following is the type of problem often included on college entrance exams where time is very valuable. Explain how you can find the answer quickly. Simplify the expression. Assume all variables are nonzero. $\left(\dfrac{[x^2y^3z]}{[x^{-7}y^{-4}]} \cdot \dfrac{1}{[y^{-2}z^{-4}]} \right)^0$
8.3	The sales s of a retail store for year t can be modeled by $s = 5423(1.12)^t$, where $t = 0$ corresponds to 1996. Find the ratio of 2001 sales to 1998 sales.
8.4	Evaluate the expression without using a calculator. Write the result in decimal form. $(5 \times 10^7) \cdot (9 \times 10^{-3})$
8.5	Find the balance after 5 years of an account that pays 3.5% interest compounded yearly, for an initial deposit of $600.
8.6	Find the value of a $15,000 car after 4 years if the car depreciates 12% per year.

Home Involvement Activity

You will need: A sheet of scrap paper, pencil, paper

Directions: Tear the scrap paper in half and throw one half away. Tear the remaining piece in half and throw half of it away. What fraction of the original paper is left? Continue tearing. Record the fraction of the original paper left after each tear. Write an exponential expression for the amount of paper left after t tears. Theoretically, will all the paper ever be thrown away? Discuss the relationship between this experiment and the decay of a radioactive material with a short half-life.

Activity: $\dfrac{1}{4}$; $\left(\dfrac{1}{2}\right)^t$ or $\left(1 - \dfrac{1}{2}\right)^t$; no

8.3: about 1.4 to 1 **8.4:** 450,000 **8.5:** about \$712.61 **8.6:** about \$8,995

8.1: second offer; the heir gets \$1,048,575 with the second offer **8.2:** notice the zero exponent first; 1

Answers

NAME _____   DATE _____

Prerequisite Skills Review

For use before Chapter 8

EXAMPLE 1 *Writing Powers*

Write the expression in exponential form.

Thirty-nine squared

SOLUTION

The base is 39. The exponent is 2.

The expression is written 39^2.

Exercises for Example 1

Write the expression in exponential form.

1. One half squared

2. Negative three cubed

3. $5x \cdot 5x \cdot 5x$

4. Eighty-two squared

EXAMPLE 2 *Exponents and Grouping Symbols*

Evaluate the expression.

$\dfrac{99}{y^2}$ when $y = -9$

SOLUTION

$\dfrac{99}{y^2} = \dfrac{99}{(-9)^2}$ Substitute -9 in for y.

$= \dfrac{99}{81}$ Evaluate power.

$= \dfrac{11}{9}$ Simplify.

Exercises for Example 2

Evaluate the expression.

5. $2a^6$ when $a = -1$

6. $m(-m^3)$ when $m = 4$

7. $\dfrac{36}{3x^4}$ when $x = 2$

8. $27r^5$ when $r = \dfrac{1}{3}$

NAME _____ DATE _____

Prerequisite Skills Review

For use before Chapter 8

EXAMPLE 3 *Solving Equations*

Solve the equation.

$$-\tfrac{3}{5}(x + 6) = 30$$

SOLUTION

$$-\tfrac{3}{5}(x + 6) = 30$$ Write original equation.

$$\left(-\tfrac{5}{3}\right)\left(-\tfrac{3}{5}\right)(x + 6) = \left(-\tfrac{5}{3}\right)(30)$$ Multiply by reciprical $-\tfrac{5}{3}$.

$$x + 6 = -50$$ Simplify.

$$x = -56$$ Subtract 6 from each side.

Exercises for Example 3

Solve the equation.

9. $12(x + 3) = 4(x + 5)$ **10.** $18(x + 2) = 15(x + 2)$

11. $x - \tfrac{1}{6} = \tfrac{5}{6}(x - 10)$ **12.** $4x - 12 = \tfrac{1}{3}(51 - 17x)$

EXAMPLE 4 *Graphing Using Slope-Intercept Form*

Graph the equation.

$$3x + y = -8$$

SOLUTION

Write the equation in slope-intercept form.

Find the slope and the y-intercept. $y = -3x - 8$

Plot the point $(0, -8)$. $m = 3, b = -8$

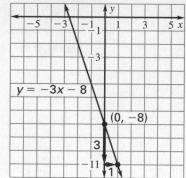

Draw a slope triangle to locate a second point on the line.

$$m = \frac{-3}{1} = \frac{rise}{run}$$

Draw a line through the points.

Exercises for Example 4

Graph the equation.

13. $y = 2x + 12$ **14.** $y = \tfrac{2}{3}x + 6$

15. $y = -\tfrac{1}{4}x + 5$ **16.** $y = 5x$

Algebra 1
Chapter 8 Resource Book

NAME _____ DATE _____

Strategies for Reading Mathematics

For use with Chapter 8

Strategy: Reading Properties and Rules

Symbols are used when writing algebraic properties and rules in order to make the ideas clearer and easier to see and to remember. Reading these properties is not hard if you take it step by step and think about what each symbol stands for.

The example below shows steps you can follow to read and understand a property involving exponents. Remember that exponential notation is a shorthand way of writing repeated multiplication. For example, to say that a variable x is taken as a factor six times, you can just write the following symbols.

$$\text{base} \rightarrow x^6 \leftarrow \text{exponent} \qquad (\text{Read as "}x\text{ to the sixth power."})$$

Product of Powers Property

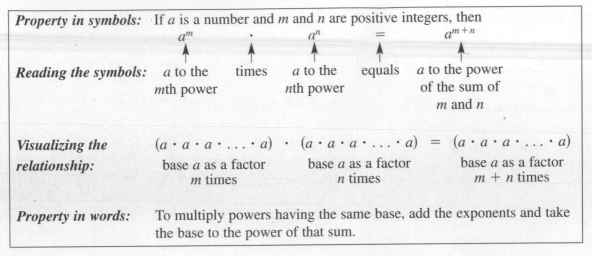

Property in symbols: If a is a number and m and n are positive integers, then

$$a^m \cdot a^n = a^{m+n}$$

Reading the symbols: a to the mth power / times / a to the nth power / equals / a to the power of the sum of m and n

Visualizing the relationship:

$$(a \cdot a \cdot a \cdot \ldots \cdot a) \cdot (a \cdot a \cdot a \cdot \ldots \cdot a) = (a \cdot a \cdot a \cdot \ldots \cdot a)$$

base a as a factor m times / base a as a factor n times / base a as a factor $m + n$ times

Property in words: To multiply powers having the same base, add the exponents and take the base to the power of that sum.

STUDY TIP
Reading Properties

Read the property part by part. Use a drawing, numerical example, or other representation to visualize what the property is saying. Try to express the ideas of the property in your own words.

STUDY TIP
Using Resources

If you forget what a symbol means, find the table of symbols in your textbook. If you forget what a property is all about, use the glossary or use the index to find pages on which the property is taught or applied.

Questions

1. Identify the exponent and the base of the power that appears in the function $y = ab^x$. Then use arrows and labels to show how to read this function.

2. The distributive property for addition can be written $a(b + c) = ab + ac$. Read and record in words what the symbols say. Next use a numerical example or a diagram to visualize the relationship. Then explain the property in your own words.

3. The formula for the volume of a cube is $V = s^3$. Follow the three steps shown above to interpret this formula. Tell what each variable stands for.

NAME _____ DATE _____

Strategies for Reading Mathematics

For use with Chapter 8

Visual Glossary

The Study Guide on page 448 lists the key vocabulary for Chapter 8 as well as review vocabulary from previous chapters. Use the page references on page 448 or the Glossary in the textbook to review key terms from prior chapters. Use the visual glossary below to help you understand some of the key vocabulary in Chapter 8. You may want to copy these diagrams into your notebook and refer to them as you complete the chapter.

GLOSSARY

exponential growth (p. 477) A quantity that is increasing by the same percent in each unit of time.

growth factor (p. 477) The expression $1 + r$ in the exponential growth model where r is the growth rate.

initial amount (pp. 477, 484) The variable C in the exponential growth or decay model.

decay factor (p. 484) The expression $1 - r$ in the exponential decay model where r is the decay rate.

exponential decay (p. 484) A quantity that is decreasing by the same percent in each unit of time.

Modeling Exponential Growth

Using the formula to plot data points and connecting the points with a curve shows the increase over time.

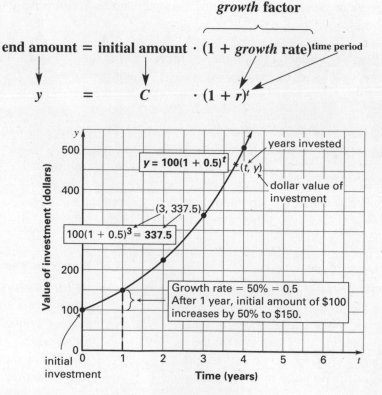

$$\text{end amount} = \text{initial amount} \cdot (1 + \textit{growth rate})^{\text{time period}}$$

$$y \quad = \quad C \quad \cdot (1 + r)^t$$

Modeling Exponential Decay

Just as exponential growth shows an increase over time, exponential decay shows a decrease over time.

$$\text{end amount} = \text{initial amount} \cdot (1 - \textit{decay rate})^{\text{time period}}$$

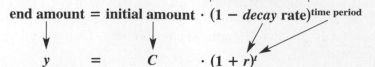

$$y \quad = \quad C \quad \cdot (1 + r)^t$$

TEACHER'S NAME _____ CLASS _____ ROOM _____ DATE _____

Lesson Plan

2-day lesson (See *Pacing the Chapter,* TE pages 446C–446D)   **For use with pages 449–455**

GOALS 1. **Use properties of exponents to multiply exponential expressions.**
2. **Use powers to model real-life problems.**

State/Local Objectives _____

_____

✓ Check the items you wish to use for this lesson.

STARTING OPTIONS
____ Prerequisite Skills Review: CRB pages 5–6
____ Strategies for Reading Mathematics: CRB pages 7–8
____ Warm-Up or Daily Homework Quiz: TE pages 450 and 437, CRB page 11, or Transparencies

TEACHING OPTIONS
____ Motivating the Lesson: TE page 451
____ Concept Activity: SE page 449; CRB page 12 (Activity Support Master)
____ Lesson Opener (Activity): CRB page 13 or Transparencies
____ Graphing Calculator Activity with Keystrokes: CRB page 14
____ Examples: Day 1: 1–4, SE pages 450–451; Day 2: 5–6, SE page 452
____ Extra Examples: Day 1: TE page 451 or Transp.; Day 2: TE page 452 or Transp.; Internet
____ Closure Question: TE page 452
____ Guided Practice: SE page 453; Day 1: Exs. 1–21; Day 2: none

APPLY/HOMEWORK
Homework Assignment
____ Basic   Day 1: 22–58; Day 2: 59–75, 82, 83, 90, 95, 100, 105
____ Average   Day 1: 22–58; Day 2: 59–75, 77, 82, 83, 90, 95, 100, 105
____ Advanced   Day 1: 22–58; Day 2: 59–77, 79, 82–85, 90, 95, 100, 105

Reteaching the Lesson
____ Practice Masters: CRB pages 15–17 (Level A, Level B, Level C)
____ Reteaching with Practice: CRB pages 18–19 or Practice Workbook with Examples
____ Personal Student Tutor

Extending the Lesson
____ Applications (Real-Life): CRB page 21
____ Challenge: SE page 455; CRB page 22 or Internet

ASSESSMENT OPTIONS
____ Checkpoint Exercises: Day 1: TE page 451 or Transp.; Day 2: TE page 452 or Transp.
____ Daily Homework Quiz (8.1): TE page 455, CRB page 25, or Transparencies
____ Standardized Test Practice: SE page 455; TE page 455; STP Workbook; Transparencies

Notes _____

_____

_____

_____

TEACHER'S NAME _____ CLASS _____ ROOM _____ DATE _____

Lesson Plan for Block Scheduling

1-day lesson (See *Pacing the Chapter,* TE pages 446C–446D)   **For use with pages 449–455**

 GOALS   1. **Use properties of exponents to multiply exponential expressions.**
2. **Use powers to model real-life problems.**

State/Local Objectives _____

_____

_____

✓ **Check the items you wish to use for this lesson.**

STARTING OPTIONS
_____ Prerequisite Skills Review: CRB pages 5–6
_____ Strategies for Reading Mathematics: CRB pages 7–8
_____ Warm-Up or Daily Homework Quiz: TE pages 450 and
           437, CRB page 11, or Transparencies

TEACHING OPTIONS
_____ Motivating the Lesson: TE page 451
_____ Concept Activity: SE page 449; CRB page 12 (Activity Support Master)
_____ Lesson Opener (Activity): CRB page 13 or Transparencies
_____ Graphing Calculator Activity with Keystrokes: CRB page 14
_____ Examples: Day 1: 1–4, SE pages 450–451; Day 2: 5–6, SE page 452
_____ Extra Examples: Day 1: TE page 451 or Transp.; Day 2: TE page 452 or Transp.; Internet
_____ Closure Question: TE page 452
_____ Guided Practice: SE page 453; Day 1: Exs. 1–21; Day 2: none

APPLY/HOMEWORK
Homework Assignment (See also the assignment for Lesson 8.2.)
_____ Block Schedule: Day 1: 22–58; Day 2: 59–75, 77, 82, 83, 90, 95, 100, 105, 106

Reteaching the Lesson
_____ Practice Masters: CRB pages 15–17 (Level A, Level B, Level C)
_____ Reteaching with Practice: CRB pages 18–19 or Practice Workbook with Examples
_____ Personal Student Tutor

Extending the Lesson
_____ Applications (Real-Life): CRB page 21
_____ Challenge: SE page 455; CRB page 22 or Internet

ASSESSMENT OPTIONS
_____ Checkpoint Exercises: Day 1: TE page 451 or Transp.; Day 2: TE page 452 or Transp.
_____ Daily Homework Quiz (8.1): TE page 455, CRB page 25, or Transparencies
_____ Standardized Test Practice: SE page 455; TE page 455; STP Workbook; Transparencies

Notes _____

_____

_____

CHAPTER PACING GUIDE	
Day	**Lesson**
1	Assess Ch. 7; **8.1 (begin)**
2	**8.1 (end)**; 8.2 (begin)
3	8.2 (end); 8.3 (begin)
4	8.3 (end); 8.4 (all)
5	8.5 (all); 8.6 (begin)
6	8.6 (end); Review Ch. 8
7	Assess Ch. 8; 9.1 (all)

LESSON
8.1

NAME _____ DATE _____

WARM-UP EXERCISES

For use before Lesson 8.1, pages 449–455

Evaluate when $x = 4$.

1. $x \cdot x \cdot x$

2. $(14 - x)^2$

3. $x \cdot x \cdot x^2$

4. $(-x) \cdot (-x)$

DAILY HOMEWORK QUIZ

For use after Lesson 7.6, pages 432–438

Graph the system of linear inequalities.

1. $y \le x + 1$
$x > 1$
$x < 3$
$y \ge 0$

2. $y \ge 0.5x - 1$
$y \le 0.5x + 2$
$x \ge -1$
$x \le 2$

3. Plot the points $(-1, 1)$, $(3, 4)$, and $(3, 1)$. Connect the points with segments. Write a system of linear inequalities to describe the triangular region.

NAME _____ DATE _____

Activity Support Master

For use with page 449

Product of powers	Expanded product	Number of factors	Product as a power	Sum of exponents
$7^3 \cdot 7^2$	$(7 \cdot 7 \cdot 7) \cdot (7 \cdot 7)$	5	7^5	
$2^4 \cdot 2^4$	$(2 \cdot 2 \cdot 2 \cdot 2) \cdot (2 \cdot 2 \cdot 2 \cdot 2)$	8		
$x^4 \cdot x^5$	$(x \cdot x \cdot x \cdot x) \cdot (x \cdot x \cdot x \cdot x \cdot x)$			

Pattern: _____

_____

_____

Product of a power	Expanded product	Expanded product	Number of factors	Product as a power	Product of exponents
$(5^2)^3$	$(5^2) \cdot (5^2) \cdot (5^2)$	$(5 \cdot 5) \cdot (5 \cdot 5) \cdot (5 \cdot 5)$	6	5^6	
$[(-3)^2]^2$	$[(-3)^2] \cdot [(-3)^2]$				
$(b^2)^4$					

Pattern: _____

_____

_____

Lesson 8.1

NAME _____ DATE _____

Activity Lesson Opener

For use with pages 450–455

SET UP: Work with a partner.

YOU WILL NEED: • a colored pencil

Begin with the expression in the box in the upper left hand corner, under START. Shade this box. Look at the expressions in all boxes surrounding the one you just shaded. Shade the box that contains an equivalent expression. Continue to shade boxes until you reach the box in the botton right hand corner, above FINISH.

START

8^5	$8^4 \cdot 8$	$(5 + 3)^5$	15^5
$8^5 \cdot 8$	$8^2 \cdot 8^3$	8^6	$(4 \cdot 2)^5$
$(2 \cdot 2)^5$	$8^5 \cdot 8^0$	$(4 + 4)^5$	16^5
$2 \cdot 4^5$	$(2 \cdot 4)^5$	$2^5 \cdot (4^2)^3$	8^{10}
$2^5 \cdot 4^5$	$2^5 \cdot 2^3 \cdot 2^2$	$2^5 \cdot (2 \cdot 2)^5$	6^{10}
6^{15}	$2^5 \cdot 2^5 \cdot 2^5$	6^5	8^{25}
$(2 \cdot 2 \cdot 2)^5$	4^{50}	$(2^3)^5$	2^{50}
4^{15}	$2^7 \cdot 2^8$	2^{56}	4^{15}
4^{56}	$(2 \cdot 2)^{15}$	2^{15}	$(2^7)^8$
$4^7 \cdot 4^8$	$(2^5)^{10}$	$(4^3)^5$	$2^{10} \cdot 2^5$

FINISH

Graphing Calculator Activity Keystrokes

For use with page 454

Keystrokes for Exercise 69

TI-82

2.1 [^] 3 [×] 4.4 [^] 3 [ENTER]

TI-83

2.1 [^] 3 [×] 4.4 [^] 3 [ENTER]

SHARP EL-9600c

2.1 [a^b] 3 [▶] [×] 4.4 [a^b] 3 [ENTER]

CASIO CFX-9850GA PLUS

From the main menu, choose RUN.

2.1 [^] 3 [×] 4.4 [^] 3 [EXE]

NAME _____ DATE _____

Practice A

For use with pages 450–455

Use the product of powers property to simplify the expression.

1. $2 \cdot 2$

2. $x \cdot x \cdot x$

3. $3^2 \cdot 3$

4. $n^3 \cdot n^2$

5. $m^2 \cdot m \cdot m^4$

6. $4^2 \cdot 4^5$

7. $y^6 \cdot y^3$

8. $t^3 \cdot t$

9. $2^2 \cdot 2 \cdot 2^2$

Use the power of a power property to simplify the expression.

10. $(2)^3$

11. $(-3)^3$

12. $(-1)^5$

13. $(2^2)^4$

14. $(3^4)^1$

15. $(-4^2)^2$

16. $(x^5)^2$

17. $(y^4)^3$

18. $(2x)^3$

Simplify, if possible. Write your answer as a power or as a product of powers.

19. $3^2 \cdot 3^4$

20. $(2^3)^5$

21. $x^5 \cdot x^3$

22. $(y^2)^8$

23. $(2x)^3$

24. $(-3x^4)^2$

25. $(x^2)^7$

26. $(-2x)^3(-x^2)$

27. $(xy)^3(z^6)^2$

28. $(-x^3)(-x)^2(-x)$

29. $(-y^3)(2y)^2$

30. $(5a^4) \cdot a^2$

Simplify. Then evaluate the expression when $x = 2$ and $y = 1$.

31. $x^2 \cdot y$

32. $x^3 \cdot x^2$

33. $(y^2)^3$

34. $(-x)^4 \cdot y$

35. $(x^2y^3)^2$

36. $-(x^2y^3)^4$

37. *Volume*   Find the total volume of four cubic crates identical to the one pictured below.

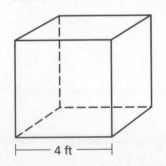

├— 4 ft —┤

38. *Volume*   The volume of a can of tuna is given by $V = \pi r^2 h$, where π is approximately 3.14. Find the volume of the can below in terms of x.

39. *Pennies*   Someone offers to double the amount of money you have every day for 1 month (30 days). You have 1 penny. On the first day, you will have 2 pennies. On the second day, you will have 4 pennies. How many pennies will you have on the 30th day? Use the expression $1 \cdot (2)^{30}$ to answer the question.

NAME _____ DATE _____

Practice B

For use with pages 450–455

Use the product of powers property to simplify the expression.

1. $4 \cdot 4 \cdot 4$

2. $n \cdot n \cdot n \cdot n$

3. $2^2 \cdot 2^3$

4. $x^2 \cdot x \cdot x^3$

5. $5^2 \cdot 5^4$

6. $c \cdot c \cdot c \cdot c^2$

7. $t^2 \cdot t^5 \cdot t$

8. $m^3 \cdot m \cdot m^4$

9. $x \cdot x^2 \cdot x^3 \cdot x^4$

Use the power of a power property to simplify the expression.

10. $(4)^2$

11. $(-5)^3$

12. $(6^2)^1$

13. $(3g)^3$

14. $(ab)^2$

15. $(ht^2)^3$

16. $(x^5)^6$

17. $(y^3)^7$

18. $(x^6y^3)^3$

Simplify, if possible. Write your answer as a power or as a product of powers.

19. $4^2 \cdot 4^3$

20. $(9^2)^5$

21. $(-4a)^2$

22. $[(-6)^2]^3$

23. $[(-6x^2y)^3]^7$

24. $[(3x - 2)^3]^3$

25. $(5x)^4 \cdot (-4x)^3$

26. $(8ab)^2 \cdot 4a^9$

27. $\left(\frac{1}{4}x^4\right)^2$

28. $(a^2bc^3)^4 \cdot (b^2c)^3$

29. $(-x)^3(-x)^5(-x)^8$

30. $(-2x^2y)(x^3y^2)^3$

Simplify. Then evaluate the expression when $x = 2$ and $y = 2$.

31. $y^2 \cdot y^4$

32. $(x^2)^2$

33. $(-x^3) \cdot x^2$

34. $(x \cdot y^3)^3$

35. $-(x^3y)^2$

36. $(y^4 \cdot y) \cdot (x)^4$

37. **Probability** A multiple choice test has two parts. There are 4^{12} ways to answer the 12 questions in Part A. There are 4^5 ways to answer the 5 questions in Part B. How many ways are there to answer all 17 questions? If you guess each answer, what is the probability you will get them all right?

38. **Nickels** Someone offers to double the amount of money you have every day for 1 month (30 days). You have 1 nickel. On the first day, you will have 2 nickels worth \$.10. On the second day, you will have 4 nickels worth \$.20. How much money will you have on the 30th day? Use the expression $0.05 \cdot (2)^{30}$ to answer the question.

39. **Maps** The scale of a square map indicates that each inch on the map corresponds to 5 miles. Write an expression that describes the area of land shown on the map. If the map is 8 inches on one side, what is the area of land shown on the map?

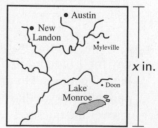

Algebra 1
Chapter 8 Resource Book

NAME _____ DATE _____

Practice C

For use with pages 450–455

Use the product of powers property to simplify the expression.

1. $x \cdot x \cdot x \cdot x \cdot x$

2. $3^3 \cdot 3^2$

3. $y^7 \cdot y \cdot y^2$

4. $z^9 \cdot z^3 \cdot z^5$

5. $6^4 \cdot 6^6 \cdot 6^1$

6. $t^3 \cdot t^3 \cdot t^3$

Use the power of a power property to simplify the expression.

7. $(4x)^2$

8. $(5x^2)^2$

9. $(2t^2)^3$

10. $(m^2 \cdot n^5)^2$

11. $(-2w^3)^4$

12. $(-3y^2)^3$

Simplify, if possible. Write your answer as a power or as a product of powers.

13. $(2)^3(2)^5$

14. $(8^3)^2$

15. $(-2x^2y^3)^2$

16. $(-3a^2c) \cdot (3b^3c^7)^4$

17. $\left(\frac{1}{2}x\right)^3$

18. $\left(-\frac{1}{3}x^4\right)^2$

19. $(3x^3)^4\left(\frac{1}{4}x^3\right)^2$

20. $(4y)^2(-3y^2)^3$

21. $[(9x + 15)^3]^6$

22. $[(-2x^4)^3(-x^8)]^2$

23. $-(a^7b^2) \cdot (a^4b^9)^3$

24. $(r^3s^7t^5)^3(s^2t)^5$

Simplify. Then evaluate the expression when $x = 2$ and $y = 1$.

25. $(x^4y^2)(y^5)$

26. $(-2xy)^3$

27. $\left(-\frac{2}{3}x\right)^2\left(\frac{3}{2}y\right)^3$

28. $(xy^2)^2(5y^3)$

29. $(2y)^4(3y^2)^2$

30. $(-3x)^3(4y^3)^2$

31. *Quarters*   Someone offers to double the amount of money you have every day for 20 days. You have 1 quarter. On the first day, you will have 2 quarters worth $.50. On the second day, you will have 4 quarters worth $1.00. How much money will you have on the 20th day?

Probability   In Exercises 32–35, use the following information.

Part A of your history test has 15 multiple choice questions.
Each question has 4 choices. Part B has 10 true/false questions.

32. How many ways are there to answer the 15 multiple choice questions?

33. How many ways are there to answer the 10 true/false questions?

34. How many ways are there to answer all 25 questions?

35. If you guess the answer to each question, what is the probability that you will get them all right?

NAME _____ DATE _____

Reteaching with Practice

For use with pages 450–455

GOAL Use properties of exponents to multiply exponential expressions and use powers to model real-life problems

---

VOCABULARY

Let a and b be numbers and let m and n be positive integers.

Product of Powers Property
To multiply powers having the same base, add the exponents.
$a^m \cdot a^n = a^{m+n}$ Example: $3^2 \cdot 3^7 = 3^{2+7} = 3^9$

Power of a Power Property
To find a power of a power, multiply the exponents.
$(a^m)^n = a^{m \cdot n}$ Example: $(5^2)^4 = 5^{2 \cdot 4} = 5^8$

Power of a Product Property
To find a power of a product, find the power of each factor and multiply.
$(a \cdot b)^m = a^m \cdot b^m$ Example: $(2 \cdot 3)^6 = 2^6 \cdot 3^6$

EXAMPLE 1 *Using the Product of Powers Property*

a. $4^3 \cdot 4^5$ **b.** $(-x)(-x)^2$

SOLUTION

To multiply powers having the same base, add the exponents.

a. $4^3 \cdot 4^5 = 4^{3+5}$ **b.** $(-x)(-x)^2 = (-x)^1(-x)^2$
$\qquad\quad = 4^8$ $\qquad\qquad\qquad = (-x)^{1+2}$
$\qquad\qquad\qquad\qquad\qquad\qquad\qquad = (-x)^3$

Exercises for Example 1

Use the product of powers property to simplify the expression.

1. $m \cdot m$ **2.** $6^2 \cdot 6^3$ **3.** $y^4 \cdot y^3$ **4.** $3 \cdot 3^5$

EXAMPLE 2 *Using the Power of a Power Property*

a. $(z^4)^5$ **b.** $(2^3)^2$

SOLUTION

To find a power of a power, multiply the exponents.

a. $(z^4)^5 = z^{4 \cdot 5}$ **b.** $(2^3)^2 = 2^{3 \cdot 2}$
$\qquad\quad = z^{20}$ $\qquad\qquad = 2^6$

NAME _____ DATE _____

Reteaching with Practice

For use with pages 450–455

Exercises for Example 2

Use the power of a power property to simplify the expression.

5. $(w^7)^3$ **6.** $(7^3)^5$ **7.** $(t^2)^6$ **8.** $[(-2)^3]^2$

EXAMPLE 3 *Using the Power of a Product Property*

Simplify $(-4mn)^2$.

SOLUTION

To find a power of a product, find the power of each factor and multiply.

$$(-4mn)^2 = (-4 \cdot m \cdot n)^2 \qquad \text{Identify factors.}$$
$$= (-4)^2 \cdot m^2 \cdot n^2 \qquad \text{Raise each factor to a power.}$$
$$= 16m^2n^2 \qquad \text{Simplify.}$$

Exercises for Example 3

Use the power of a product property to simplify the expression.

9. $(5x)^3$ **10.** $(10s)^2$ **11.** $(-x)^4$ **12.** $(-3y)^3$

EXAMPLE 4 *Using Powers to Model Real-Life Problems*

You are planting two square vegetable gardens. The side of the larger garden is twice as long as the side of the smaller garden. Find the ratio of the area of the larger garden to the area of the smaller garden.

SOLUTION

$$\text{Ratio} = \frac{(2x)^2}{x^2} = \frac{2^2 \cdot x^2}{x^2} = \frac{4x^2}{x^2} = \frac{4}{1}$$

Exercise for Example 4

13. Rework Example 4 if the side of the larger garden is three times as long as the side of the smaller garden.

NAME _____ DATE _____

Quick Catch-Up for Absent Students

For use with pages 449–455

The items checked below were covered in class on (date missed) _____

Activity 8.1: Investigating Powers (p. 449)

_____ **Goal:** Multiply exponential expressions using addition and multiplication.

Lesson 8.1: Multiplication Properties of Exponents

_____ **Goal 1:** Use properties of exponents to multiply exponential expressions. (pp. 450–451)

Material Covered:

_____ Student Help: Look Back

_____ Example 1: Using the Product of Powers Property

_____ Example 2: Using the Power of a Power Property

_____ Student Help: Study Tip

_____ Example 3: Using the Power of a Product Property

_____ Example 4: Using All Three Properties

_____ **Goal 2:** Use powers to model real-life problems. (p. 452)

Material Covered:

_____ Example 5: Using the Power of a Product Property

_____ Student Help: Skills Review

_____ Example 3: Using the Product of Powers Property

_____ Other (specify) _____

_____

Homework and Additional Learning Support

_____ Textbook (specify) <u>pp. 453–455</u>_____

_____

_____ Internet: Extra Examples at www.mcdougallittel.com

_____ *Reteaching with Practice* worksheet (specify exercises)_____

_____ *Personal Student Tutor* for Lesson 8.1

Real-Life Application:
When Will I Ever Use This?

For use with pages 450–455

Telephone Numbers

Did you ever wonder how you got your telephone number? The North American Numbering Plan (NANP) was invented in 1947. Under the plan, a phone number contained 10 digits. The first three digits were the area code, the second three digits were the exchange, and the last four digits were the individual telephone line numbers. Under the original plan, the first number of an area code could not be a 0 or a 1, and the second number had to be a 0 or a 1. Similarly, the first number of the exchange could not be a 0 or 1. This system created 160 possible area codes, of which 144 were assigned throughout North America. Sixteen of the area codes were reserved for special uses, such as emergency and toll-free numbers.

In 1992 the last of the original area codes was assigned, prompting a need for additional area codes. To alleviate the problem, a new area code format was introduced that removed the restriction on the second number of the area code. This adjustment created 640 additional area code possibilities. The demand for new numbers increases daily with the use of phone lines for business lines, faxes, computer modems, cellular phones, and pagers.

The assignment of new area codes has become common across the country. The two most popular methods for assigning new area codes are the overlay and the geographic split. An overlay just adds the new area code to an existing geographical area without changing its original area code. New customers are assigned the new area code and existing customers retain their original number. Although this method is often the least hassle for individual customers and businesses, an overlay will require all customers to dial ten digits for all calls, even local ones. A geographic split, on the other hand, just divides a geographic area into two regions. One section keeps the same area code, and the other receives a new one.

**In Exercises 1 and 2, use the information above to answer
the question.**

   1. Using the new NANP guidelines, write an expression in exponential form.

      a. the possible number of area codes

      b. the possible number of exchanges

      c. the possible number of individual line numbers

   2. Use your answers in Exercise 1 and the multiplication properties of exponents to find each amount. Show your work.

      a. the number of possible phone numbers in one area code

      b. the number of possible 10-digit phone numbers

Challenge: Skills and Applications

For use with pages 450–455

In Exercises 1–6, simplify if possible. Write your answer as a power.

1. $k^n \cdot k^3$

2. $x^{4n+5} \cdot x^{2n-3}$

3. $(y^{7m})^3$

4. $(k^{4m})^{3n}$

5. $(x^3 y^m)^{2n}$

6. $(x^{4m} y^{6m})^{9n} \cdot x^7$

In Exercises 7–8, use the box shown.

7. Write an expression that gives the volume of the box in terms of a and b.

8. Suppose a is halved and b is multiplied by 8. By what factor is the volume of the box multiplied?

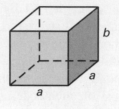

In Exercises 9–16, for each pair of nonnegative numbers a and b and each positive integer n, calculate the value of $a^n + b^n$ and the value of $(a + b)^n$.

9. $a = 3, b = 4, n = 2$

10. $a = 2, b = 5, n = 3$

11. $a = 4, b = 0, n = 4$

12. $a = 1, b = 1, n = 7$

13. $a = 6, b = 2, n = 1$

14. $a = 3, b = 3, n = 3$

15. $a = 0, b = 10, n = 5$

16. $a = 9, b = 2, n = 1$

17. Based on your answers from Exercises 9–16, under what conditions, if any, on a, b, and/or n, is it true that $a^n + b^n = (a + b)^n$?

18. A test has 8 true-false and 12 multiple choice questions. Each multiple choice question has 4 possible answers. Write an exponential expression for the number of different ways it is possible to answer the 20 questions.

19. Use the fact that $4 = 2^2$ to write the expression from Exercise 18 as a power of 2.

TEACHER'S NAME _____ CLASS _____ ROOM _____ DATE _____

Lesson Plan

2-day lesson (See *Pacing the Chapter,* TE pages 446C–446D)   For use with pages 456–462

GOALS
1. **Evaluate powers that have zero and negative exponents.**
2. **Graph exponential functions.**

State/Local Objectives _____

_____

✓ **Check the items you wish to use for this lesson.**

STARTING OPTIONS

____ Homework Check: TE page 453; Answer Transparencies
____ Warm-Up or Daily Homework Quiz: TE pages 456 and 455, CRB page 25, or Transparencies

TEACHING OPTIONS

____ Motivating the Lesson: TE page 457
____ Lesson Opener (Visual Approach): CRB page 26 or Transparencies
____ Graphing Calculator Activity with Keystrokes: CRB pages 27–28
____ Examples: Day 1: 1–4, SE pages 456–457; Day 2: 5–6, SE page 458
____ Extra Examples: Day 1: TE page 457 or Transp.; Day 2: TE page 458 or Transp.; Internet
____ Technology Activity: SE page 462
____ Closure Question: TE page 458
____ Guided Practice: SE page 459; Day 1: Exs. 1–10; Day 2: Exs. 11–13

APPLY/HOMEWORK

Homework Assignment

____ Basic   Day 1: 14–44 even, 46–48, 50–60 even; Day 2: 15–45 odd, 51–61 odd, 66–69, 71–75
____ Average   Day 1: 14–44 even, 46–48, 50–62 even; Day 2: 15–45 odd, 51–63 odd, 66–69, 71–75
____ Advanced   Day 1: 14–44 even, 46–48, 50–64 even; Day 2: 15–45 odd, 51–63 odd, 66–69, 70–75

Reteaching the Lesson

____ Practice Masters: CRB pages 29–31 (Level A, Level B, Level C)
____ Reteaching with Practice: CRB pages 32–33 or Practice Workbook with Examples
____ Personal Student Tutor

Extending the Lesson

____ Applications (Interdisciplinary): CRB page 35
____ Challenge: SE page 461; CRB page 36 or Internet

ASSESSMENT OPTIONS

____ Checkpoint Exercises: Day 1: TE page 457 or Transp.; Day 2: TE page 458 or Transp.
____ Daily Homework Quiz (8.2): TE page 461, CRB page 39, or Transparencies
____ Standardized Test Practice: SE page 461; TE page 461; STP Workbook; Transparencies

Notes _____

_____

_____

TEACHER'S NAME _____ CLASS _____ ROOM _____ DATE _____

Lesson Plan for Block Scheduling

1-day lesson (See *Pacing the Chapter,* **TE pages 446C–446D)**   **For use with pages 456–462**

 GOALS   1. **Evaluate powers that have zero and negative exponents.**
   2. **Graph exponential functions.**

State/Local Objectives _____

_____

_____

✓ **Check the items you wish to use for this lesson.**

STARTING OPTIONS

____ Homework Check: TE page 453; Answer Transparencies

____ Warm-Up or Daily Homework Quiz: TE pages 456 and
        455, CRB page 25, or Transparencies

TEACHING OPTIONS

____ Motivating the Lesson: TE page 457

____ Lesson Opener (Visual Approach): CRB page 26 or Transparencies

____ Graphing Calculator Activity with Keystrokes: CRB pages 27–28

____ Examples: Day 2: 1–4, SE pages 456–457; Day 3: 5–6, SE page 458

____ Extra Examples: Day 2: TE page 457 or Transp.; Day 3: TE page 458 or Transp.; Internet

____ Technology Activity: SE page 462

____ Closure Question: TE page 458

____ Guided Practice: SE page 459; Day 2: Exs. 1–10; Day 3: Exs. 11–13

APPLY/HOMEWORK

Homework Assignment (See also the assignments for Lessons 8.1 and 8.3.)

____ Block Schedule: Day 2: 14–44 even, 46–48, 50–60 even, 64;
                        Day 3: 15–45 odd, 51–63 odd, 66–69, 71–75

Reteaching the Lesson

____ Practice Masters: CRB pages 29–31 (Level A, Level B, Level C)

____ Reteaching with Practice: CRB pages 32–33 or Practice Workbook with Examples

____ Personal Student Tutor

Extending the Lesson

____ Applications (Interdisciplinary): CRB page 35

____ Challenge: SE page 461; CRB page 36 or Internet

ASSESSMENT OPTIONS

____ Checkpoint Exercises: Day 2: TE page 457 or Transp.; Day 3: TE page 458 or Transp.

____ Daily Homework Quiz (8.2): TE page 461, CRB page 39, or Transparencies

____ Standardized Test Practice: SE page 461; TE page 461; STP Workbook; Transparencies

Notes _____

_____

_____

| CHAPTER PACING GUIDE ||
Day	Lesson
1	Assess Ch. 7; 8.1 (begin)
2	8.1 (end); **8.2 (begin)**
3	**8.2 (end)**; 8.3 (begin)
4	8.3 (end); 8.4 (all)
5	8.5 (all); 8.6 (begin)
6	8.6 (end); Review Ch. 8
7	Assess Ch. 8; 9.1 (all)

Lesson 8.2

NAME _____ DATE _____

WARM-UP EXERCISES

For use before Lesson 8.2, pages 456–462

Evaluate each expression.

1. 4^2

2. $(-2)^3$

3. $\left(\dfrac{1}{2}\right)^3$

4. $2 \cdot 1.05^2$

5. $3^4 \cdot 3^3$

DAILY HOMEWORK QUIZ

For use after Lesson 8.1, pages 449–455

Simplify. Write your answer as a power or as a product of powers.

1. $5^7 \cdot 5^4$

2. $(3^4)^5$

3. $(7m)^6$

4. $(-4x^7y^2)^3$

5. $[(5p + 2)^2]^5$

6. $\left(\dfrac{2}{3}h^4\right)^3(5h^2)^4$

7. $(-6s^3t)^2(9s^2t^4)^3$

8. Simplify $(-b)^4 \cdot a^3 \cdot (b \cdot a)$. Evaluate the resulting expression when $a = -3$ and $b = -2$.

9. Complete the statement using $>$ or $<$.
$(4^3 \cdot 5^2)^2$ _____ $4^9 \cdot 5^4$

NAME _____ DATE _____

Visual Approach Lesson Opener

For use with pages 456–461

1. The function $y = 2^x$ is graphed below. Find the coordinates of the points shown on the graph.

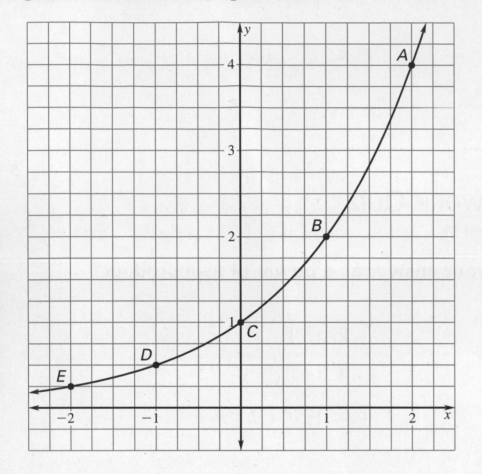

2. What do you notice about the value of y when x is positive?

3. What do you notice about the value of y when x is 0?

4. What do you notice about the value of y when x is negative?

NAME _____ DATE _____

Graphing Calculator Activity Keystrokes

For use with pages 457

Keystrokes for Example 3

TI-82

3 [^] [(-)] 4 [ENTER]

TI-83

3 [^] [(-)] 4 [ENTER]

SHARP EL-9600c

3 [a^b] [(-)] 4 [ENTER]

CASIO CFX-9850GA PLUS

From the main menu, choose RUN.

3 [^] [(-)] 4 [EXE]

Graphing Calculator Activity Keystrokes

For use with Technology Activity 8.2 on page 462

TI-82

Y= 0.5 ^ X,T,θ ENTER

WINDOW ENTER

(-) 10 ENTER

10 ENTER

2 ENTER

(-) 10 ENTER

10 ENTER

2 ENTER

GRAPH

TI-83

Y= 0.5 ^ X,T,θ,n ENTER

WINDOW

(-) 10 ENTER

10 ENTER

2 ENTER

(-) 10 ENTER

10 ENTER

2 ENTER

GRAPH

SHARP EL-9600c

Y= 0.5 a^b X/θ/T/n ENTER

WINDOW

(-) 10 ENTER

10 ENTER

2 ENTER

(-) 10 ENTER

10 ENTER

2 ENTER

GRAPH

CASIO CFX-9850GA PLUS

From the main menu, choose GRAPH.

0.5 ^ X,θ,T EXE

SHIFT F3 (-) 10 EXE 10 EXE 2 EXE

(-) 10 EXE 10 EXE 2 EXE

EXIT F6

Practice A

For use with pages 456–461

Complete the table.

	Exponent, n	3	2	1	0	-1	-2	-3
1.	Power, 2^n	8						
2.	Power, 3^n	27						
3.	Power, 4^n	64						

Evaluate the exponential expression. Write your answer as a fraction in simplest form.

4. 3^{-3}

5. 2^{-5}

6. 5^0

7. $8^0 \cdot 2^{-3}$

8. $3^5 \cdot 3^{-4}$

9. $5^{-7} \cdot 5^9$

10. $(2^3)^2$

11. $(6^{-1})^2$

12. $(-2^3)^{-1}$

Rewrite the expression with positive exponents.

13. x^{-8}

14. $3x^{-5}$

15. $\dfrac{7}{x^{-2}}$

16. $\dfrac{9}{x^{-4}}$

17. $8x^{-7}y^{-8}$

18. $3a^{-3}$

19. $\dfrac{3x^0}{y^{-3}}$

20. $(4x)^{-2}$

21. $(-2x)^{-4}$

22. $(5x)^0 y^{-2}$

23. $\dfrac{1}{(3x)^{-3}}$

24. $(2x)^{-2} \cdot 3y^5$

25. Complete the table.

x	-3	-2	-1	0	1	2	3
$y = 3^x$							

26. Graph the table of values you found in Exercise 25.

27. For the graph in Exercise 26, as the value of x increases, what happens to the value of y?

28. Complete the table.

x	-3	-2	-1	0	1	2	3
$y = \left(\frac{1}{2}\right)^x$							

29. Graph the table of values you found in Exercise 28.

30. For the graph in Exercise 29, as the value of x increases, what happens to the value of y?

NAME _____ DATE _____

Practice B

For use with pages 456–461

Evaluate the exponential expression. Write your answer as a fraction in simplest form.

1. 5^{-3}

2. $\left(\frac{1}{3}\right)^{-1}$

3. $6(6^{-4})$

4. $-2^0 \cdot \dfrac{1}{4^{-2}}$

5. $3^5 \cdot 3^{-7}$

6. $7^3 \cdot 0^{-2}$

7. $10^{-2} \cdot 10^2$

8. $-2 \cdot (-2)^{-5}$

9. $(8^2)^{-1}$

10. $9^{-2} \cdot 12^0$

11. $(-4^{-3})^{-1}$

12. $1 \cdot 1^{-8}$

Rewrite the expression with positive exponents.

13. $4x^{-2}$

14. $\dfrac{1}{3x^{-4}}$

15. $x^3 y^{-6}$

16. $7x^{-5}y^{-1}$

17. $\dfrac{1}{11x^{-2}y^{-7}}$

18. $(-12)^0 y^{-2}$

19. $(9x)^{-4}$

20. $(2x^3 y^{-8})^{-3}$

21. $(2^{-1}x^{-10})^7$

22. $\dfrac{15}{5y^{-3}}$

23. $\dfrac{1}{(8x^2)^{-3}}$

24. $\left(\dfrac{-12x^{-5}}{4x^{-5}}\right)^{-4}$

25. Complete the table.

x	-2	-1	0	1	2
$y = \left(\frac{2}{3}\right)^x$					

26. Graph the table of values in Exercise 25.

27. For the graph in Exercise 26, as the value of x increases, what happens to the value of y?

28. **Endangered Species**   Between 1990 and 2000, the population of an endangered species decreased at a rate of 0.1% per year. The population P in year t is given by $P = 1200(0.999)^t$, where $t = 0$ corresponds to 1995. Find the population of the species in 1990, 1995, 2000, and the projected population in 2010.

	1990 $(t = -5)$	1995 $(t = 0)$	2000 $(t = 5)$	2010 $(t = 15)$
t				
$P = 1200(0.999)^t$				

29. **Town Population**   Between 1960 and 1990, the population of a town increased at a rate of 0.34% per year. The population P in year t is given by $P = 2000(1.0034)^t$, where $t = 0$ corresponds to 1980. Find the population of the town in 1960, 1970, 1980, and 1990.

	1960 $(t = -20)$	1970 $(t = -10)$	1980 $(t = 0)$	1990 $(t = 10)$
t				
$P = 2000(1.0034)^t$				

NAME _____ DATE _____

Practice C

For use with pages 456–461

Evaluate the exponential expression. Write your answer as a fraction in simplest form.

1. 12^{-2}

2. $\left(\frac{2}{5}\right)^{-3}$

3. $8^5(8^{-7})$

4. $(-10)^0 \cdot \dfrac{1}{3^{-3}}$

5. $6^{13} \cdot 6^{-10}$

6. $11^{-2} \cdot 0^{-6}$

7. $21^{-8} \cdot 21^8$

8. $-9 \cdot (-9)^{-3}$

9. $(5^3)^{-1}$

10. $10^{-3} \cdot 20^0$

11. $(-3^{-1})^{-5}$

12. $15^{-5} \cdot 0^9$

Rewrite the expression with positive exponents.

13. $14x^{-5}$

14. $\dfrac{4}{5^{-2}x^{-7}}$

15. $x^{-10}y^{21}$

16. $20x^{-8}y^{-8}$

17. $\dfrac{6}{18x^{-3}y^9}$

18. $(-11)^{-2}y^0$

19. $(7^{-2}x^8)^{-2}$

20. $(4x^{-4}y^{-12})^{-5}$

21. $\dfrac{-48x^{-6}y^4}{52x^9y^{-2}}$

22. $\dfrac{8^{-2}}{2^{-4}x^{-4}}$

23. $\dfrac{x^{-4}}{(12y^2)^{-2}}$

24. $\left(\dfrac{-10x^{-15}}{x^{-15}}\right)^{-5}$

25. Complete the table.

x	-2	-1	0	1	2
$y = (0.40)^x$					

26. Graph the table of values in Exercise 25.

27. For the graph in Exercise 26, as the value of x increases, what happens to the value of y?

28. *Endangered Species*   Between 1990 and 2000, the population of an endangered species decreased at a rate of 0.1% per year. The population P in year t is given by $P = 1200(0.999)^t$, where $t = 0$ corresponds to 1995. Find the population of the species in 1990, 1995, 2000, and the projected population in 2010.

29. *Town Population*   Between 1960 and 1990, the population of a town increased at a rate of 0.34% per year. The population P in year t is given by $P = 2000(1.0034)^t$, where $t = 0$ corresponds to 1980. Find the population of the town in 1960, 1970, 1980, and 1990.

30. *Radium Isotope*   The half-life of the radium isotope Ra^{226} is about 1620 years. If there were initially 100 grams of Ra^{226}, then the number of grams remaining after h half-life periods is $W = 100\left(\frac{1}{2}\right)^h$. Complete the table.

Half-life period, h	0	1	2	3	4	5	6
Grams, W							

31. *Savings Account*   You started a savings account in 1990. The balance A is modeled by $A = 600(1.07)^t$, where $t = 0$ represents the year 2000. What is the balance in the account in 1990? in 2000? in 2010?

NAME _____   DATE _____

Reteaching with Practice

For use with pages 456–461

GOAL Evaluate powers that have zero and negative exponents and graph exponential functions

---

VOCABULARY

Let a be a nonzero number and let n be a positive integer.

- A nonzero number to the zero power is 1: $a^0 = 1$, $a \neq 0$.

- a^{-n} is the reciprocal of a^n: $a^{-n} = \dfrac{1}{a^n}$, $a \neq 0$.

An **exponential function** is a function of the form $y = ab^x$, where $b > 0$ and $b \neq 1$.

EXAMPLE 1 *Powers with Zero and Negative Exponents*

Evaluate the exponential expression. Write your answer as a fraction in simplest form.

a. $(-8)^0$ **b.** 4^{-2}

SOLUTION

a. $(-8)^0 = 1$ A nonzero number to the zero power is 1.

b. $4^{-2} = \dfrac{1}{4^2} = \dfrac{1}{16}$ 4^{-2} is the reciprocal of 4^2.

Exercises for Example 1

Evaluate the exponential expression. Write your answer as a fraction in simplest form.

1. 73^0 **2.** $\left(\tfrac{1}{2}\right)^{-1}$ **3.** 13^{-x}

EXAMPLE 2 *Simplifying Exponential Expressions*

Rewrite the expression with positive exponents.

a. $5y^{-1}z^{-2}$ **b.** $(2x)^{-3}$

SOLUTION

a. $5y^{-1}z^{-2} = 5 \cdot \dfrac{1}{y} \cdot \dfrac{1}{z^2} = \dfrac{5}{yz^2}$

b. $(2x)^{-3} = 2^{-3} \cdot x^{-3}$ Use power of a product property.

        $= \dfrac{1}{2^3} \cdot \dfrac{1}{x^3}$ Write reciprocals of 2^3 and x^3.

        $= \dfrac{1}{8x^3}$ Multiply fractions.

NAME _____ DATE _____

Reteaching with Practice

For use with pages 456–461

Exercises for Example 2

Rewrite the expression with positive exponents.

4. $(13y)^{-1}$ **5.** $\dfrac{1}{(2x)^{-4}}$ **6.** $(2c)^{-4}d$

EXAMPLE 3   *Evaluating Exponential Expressions*

Evaluate the expression.

$(3^{-2})^{-3}$

SOLUTION

$(3^{-2})^{-3} = 3^{-2 \cdot (-3)}$ Use power of a power property.

            $= 3^6$ Multiply exponents.

            $= 729$ Evaluate.

Exercises for Example 3

Evaluate the expression.

7. $8^{-1} \cdot 8^1$ **8.** $4^6 \cdot 4^{-4}$ **9.** $(5^{-2})^2$

EXAMPLE 4   *Graphing an Exponential Function*

Sketch the graph of $y = 3^x$.

SOLUTION

Make a table that includes negative x-values.

x	-2	-1	0	1	2
3^x	$3^{-2} = \frac{1}{9}$	$3^{-1} = \frac{1}{3}$	$3^0 = 1$	3	9

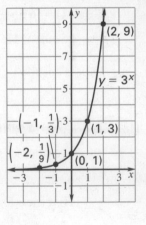

Draw a coordinate plane and plot the five points given by the table. Draw a smooth curve through the points.

Exercise for Example 4

10. Sketch the graph of $y = 4^x$.

NAME _____ DATE _____

Quick Catch-Up for Absent Students

For use with pages 456–462

The items checked below were covered in class on (date missed) _____

Lesson 8.2: Zero and Negative Exponents

____ **Goal 1:** Evaluate powers that have zero and negative exponents. (pp. 456–457)

Material Covered:

    ____ Activity: Investigating Zero and Negative Exponents

    ____ Example 1: Powers with Zero and Negative Exponents

    ____ Student Help: Study Tip

    ____ Example 2: Simplifying Exponential Expressions

    ____ Student Help: Keystroke Help

    ____ Example 3: Evaluating Exponential Expressions

    ____ Example 4: Simplifying Exponential Expressions

____ **Goal 2:** Graph exponential functions. (p. 458)

Material Covered:

    ____ Example 5: Graphing an Exponential Function

    ____ Example 6: Evaluating an Exponential Function

Vocabulary:

    exponential function, p. 458

Activity 8.2: Graphing Exponential Functions (p. 462)

____ **Goal:** Graph an exponential function using a graphing calculator.

    ____ Student Help: Keystroke Help

____ Other (specify) _____

_____

Homework and Additional Learning Support

    ____ Textbook (specify) <u>pp. 459–461</u>_____

_____

    ____ Internet: Extra Examples at www.mcdougallittel.com

    ____ *Reteaching with Practice* worksheet (specify exercises)_____

    ____ *Personal Student Tutor* for Lesson 8.2

Lesson 8.2

Interdisciplinary Application

For use with pages 456–461

Carbon-14 Dating

CHEMISTRY   Prior to World War II archeologists depended upon recorded histories for dating the past. Assuming that sites with similar types of pottery and tools were the same age, they dated sites relatively. Relative dating, however, still could not date most sites. In the late 1940s a team of scientists led by Willard Libby developed a dating method that would revolutionize the field of archeology. The importance of carbon-14 dating was recognized when Libby received the 1960 Nobel Prize in Chemistry for his work.

The discovery that all living organisms absorb a radioactive isotope of carbon called carbon-14 makes carbon-14 dating possible. When an organism dies, it stops absorbing carbon-14. The amount of this radioactive isotope then steadily decreases over time. Professor Libby found that carbon-14 has a half-life of 5730 years, or decays to half of its original amount each 5730 years. This method can only be used to date artifacts less than about 50,000 or 60,000 years.

Carbon-14 dating has been used to date such artifacts and fossils as the Dead Sea Scrolls and the famous Iceman found in Italy in 1991. With more recent developments in carbon-14 dating, it is possible to date very small samples. Even a single human hair can now be dated!

In Exercises 1-3, use the following information.

The equation for carbon-14 dating given below is based on a half-life of 5730 years.

$$A(t) = A_0(2)^{-t},$$

where $A(t)$ is the amount of carbon-14 left, A_0 is the initial amount of carbon-14, and t is the number of 5730-year intervals.

1. Make a table of values for $A(t)$ when $A_0 = 100$ grams and $t = 0, 1, 2, 3, 4$ and 5.

2. Graph the points from the table of values in Exercise 1 and draw a smooth curve through the points.

3. After considering the graph in Exercise 2, why do you think carbon-14 dating can only be used to date up to about 50,000 or 60,000 years?

Challenge: Skills and Applications

In Exercises 1–4, decide whether each expression is positive or negative for $a < 0$ and $b > 1$.

1. $-a^0$

2. $(-b)^0$

3. $a^0 b^0$

4. $a^0 - b$

In Exercises 5–8, evaluate the exponential expression. Write your answer as a whole number or a fraction in simplest form.

5. $(4^{-1} \cdot 7)^{-2}$

6. $4^{-3} \cdot \left(\frac{5}{8}\right)^{-1}$

7. $(0.01)^{-3}$

8. $(0.02)^{-2}(0.04)^3$

In Exercises 9–16, use the following information.

Banks often compute interest over time periods shorter than 1 year. For example, "6% annual interest compounded semi-annually" means that instead of growing by 6% every year, your savings grows by 3% every half-year. In the formula $y = P(1 + r)^x$, $r = 0.03$ and $x =$ the number of half-year intervals in t years. Find the value in each account, after t years, for an initial investment P of $1500 at the given annual rate I.

9. $I = 4\%$, compounded quarterly, $t = 6$

10. $I = 8\%$, compounded semi-annually, $t = 8$

11. $I = 6\%$, compounded monthly, $t = 4$

12. $I = 10\%$, compounded quarterly, $t = 5$

13. $I = 5\%$, compounded monthly, $t = 8$

14. $I = 5\%$, compounded semi-annually, $t = 8$

15. $I = 5.5\%$, compounded semi-annually, $t = 8$

16. Compare your answers from Exercises 13 and 14. How much difference does compounding make in the value of the account? Compare your answers from Exercises 14 and 15. Does the rate of interest make more difference in the final value of the account than the number of times per year that the interest is compounded?

TEACHER'S NAME _____ CLASS _____ ROOM _____ DATE _____

Lesson Plan
2-day lesson (See *Pacing the Chapter,* TE pages 446C–446D)   For use with pages 463–469

GOALS   1. **Use the division properties of exponents to evaluate powers and to simplify expressions.**
2. **Use the division property of exponents to solve real-life problems.**

State/Local Objectives _____

_____

✓ **Check the items you wish to use for this lesson.**

STARTING OPTIONS
____ Homework Check: TE page 459; Answer Transparencies
____ Warm-Up or Daily Homework Quiz: TE pages 463 and 461, CRB page 39, or Transparencies

TEACHING OPTIONS
____ Motivating the Lesson: TE page 464
____ Lesson Opener (Calculator): CRB page 40 or Transparencies
____ Graphing Calculator Activity with Keystrokes: CRB pages 41–42
____ Examples: Day 1: 1–3, SE pages 463–464; Day 2: 4–5, SE page 465
____ Extra Examples: Day 1: TE page 464 or Transp.; Day 2: TE page 465 or Transp.; Internet
____ Closure Question: TE page 465
____ Guided Practice: SE page 466; Day 1: Exs. 1–18; Day 2: none

APPLY/HOMEWORK
Homework Assignment
____ Basic   Day 1: 20–48 even; Day 2: 49–58, 62–65, 70, 75, 80, 85; Quiz 1: 1–24
____ Average   Day 1: 20–48 even; Day 2: 49–58, 62–65, 70, 75, 80, 85; Quiz 1: 1–24
____ Advanced   Day 1: 20–48 even; Day 2: 49–66, 70, 75, 80, 85; Quiz 1: 1–24

Reteaching the Lesson
____ Practice Masters: CRB pages 43–45 (Level A, Level B, Level C)
____ Reteaching with Practice: CRB pages 46–47 or Practice Workbook with Examples
____ Personal Student Tutor

Extending the Lesson
____ Applications (Real-Life): CRB page 49
____ Challenge: SE page 468; CRB page 50 or Internet

ASSESSMENT OPTIONS
____ Checkpoint Exercises: Day 1: TE page 464 or Transp.; Day 2: TE page 465 or Transp.
____ Daily Homework Quiz (8.3): TE page 469, CRB page 54, or Transparencies
____ Standardized Test Practice: SE page 468; TE page 469; STP Workbook; Transparencies
____ Quiz (8.1–8.3): SE page 469; CRB page 51

Notes _____

_____

_____

TEACHER'S NAME _____ CLASS _____ ROOM _____ DATE _____

Lesson Plan for Block Scheduling

1-day lesson (See *Pacing the Chapter,* TE pages 446C–446D)   **For use with pages 463–469**

 GOALS
1. **Use the division properties of exponents to evaluate powers and to simplify expressions.**
2. **Use the division property of exponents to solve real-life problems.**

State/Local Objectives _____

_____

_____

✓ **Check the items you wish to use for this lesson.**

STARTING OPTIONS

_____ Homework Check: TE page 459; Answer Transparencies

_____ Warm-Up or Daily Homework Quiz: TE pages 463 and
          461, CRB page 39, or Transparencies

TEACHING OPTIONS

_____ Motivating the Lesson: TE page 464

_____ Lesson Opener (Calculator): CRB page 40 or Transparencies

_____ Graphing Calculator Activity with Keystrokes: CRB pages 41–42

_____ Examples: Day 3: 1–3, SE pages 463–464; Day 4: 4–5, SE page 465

_____ Extra Examples: Day 3: TE page 464 or Transp.; Day 4: TE page 465 or Transp.; Internet

_____ Closure Question: TE page 465

_____ Guided Practice: SE page 466; Day 3: Exs. 1–18; Day 4: none

APPLY/HOMEWORK

Homework Assignment (See also the assignments for Lessons 8.2 and 8.4.)

_____ Block Schedule: Day 3: 20–48 even; Day 4: 49–58, 62–65, 70, 75, 80, 85; Quiz 1: 1–24

Reteaching the Lesson

_____ Practice Masters: CRB pages 43–45 (Level A, Level B, Level C)

_____ Reteaching with Practice: CRB pages 46–47 or Practice Workbook with Examples

_____ Personal Student Tutor

Extending the Lesson

_____ Applications (Real-Life): CRB page 49

_____ Challenge: SE page 468; CRB page 50 or Internet

ASSESSMENT OPTIONS

_____ Checkpoint Exercises: Day 3: TE page 464 or Transp.; Day 4: TE page 465 or Transp.

_____ Daily Homework Quiz (8.3): TE page 469, CRB page 54, or Transparencies

_____ Standardized Test Practice: SE page 468; TE page 469; STP Workbook; Transparencies

_____ Quiz (8.1–8.3): SE page 469; CRB page 51

Notes _____

_____

_____

CHAPTER PACING GUIDE	
Day	**Lesson**
1	Assess Ch. 7; 8.1 (begin)
2	8.1 (end); 8.2 (begin)
3	8.2 (end); **8.3 (begin)**
4	**8.3 (end)**; 8.4 (all)
5	8.5 (all); 8.6 (begin)
6	8.6 (end); Review Ch. 8
7	Assess Ch. 8; 9.1 (all)

Algebra 1
Chapter 8 Resource Book

Lesson 8.3

NAME _____ DATE _____

WARM-UP EXERCISES

For use before Lesson 8.3, pages 463–469

Simplify.

1. 5^{-2}

2. $(3w^2)^2$

3. $(4vw^3)^2$

4. $(2xy^3)^4(4x^2y^5)$

5. $(-2r^2s^3)^5$

DAILY HOMEWORK QUIZ

For use after Lesson 8.2, pages 456–462

Evaluate the exponential expression. Write fractions in simplest form.

1. $7(7^{-3})$

2. $\left(\dfrac{1}{3}\right)^{-2}$

3. $(4^{-2})^2$

4. $12 \cdot 12^{-1}$

Rewrite the expression with positive exponents.

5. $m^{-3}n^5$

6. $\dfrac{1}{7k^{-3}}$

7. $(15r)^0$

8. $\dfrac{2}{(8x)^{-3}}$

9. Graph the exponential function $y = \left(\dfrac{1}{7}\right)^x$.

Algebra 1
Chapter 8 Resource Book
39

LESSON

8.3

NAME _____ DATE _____

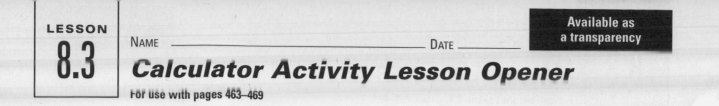

Calculator Activity Lesson Opener

For use with pages 463–469

Available as
a transparency

1. Make a table like the one below. Use the y^x key or the $\wedge$ key on your calculator to evaluate the expressions for the third and fourth columns.

a	m	n	$\dfrac{a^m}{a^n}$	$a^{(m-n)}$
2	6	3		
3	5	2		
4	4	1		

2. Compare the values in the third and fourth column for each row. What conclusion can you make? Test your conclusion by adding several more rows of values for a, m, and n and calculating.

3. Make a table like the one below. Use your calculator to evaluate the expressions for the third and fourth columns.

a	b	m	$\left(\dfrac{a}{b}\right)^m$	$\dfrac{a^m}{b^m}$
8	2	3		
12	4	2		
6	3	2		

4. Compare the values in the last two boxes of each row. What conclusion can you make? Test your conclusion by adding several more rows of values for a, b, and m and calculating.

NAME _____ DATE _____

Graphing Calculator Activity

For use with pages 463–469

GOAL **To discover the power of a quotient property**

In Lessons 8.1 and 8.2, you learned about the multiplication properties of exponents and negative and zero exponents. Your calculator can help you discover and verify the division properties of exponents.

Activity

You can use graphing to verify equivalent expressions. The graphs will appear as one curve because one expression is graphed on top of the other.

❶ Enter the expression $\left(\frac{3}{4}\right)^x$ as Y_1 in your graphing calculator. Then enter each of the following expressions into Y_2 and plot the graphs in the same coordinate plane. Which expression is equivalent to the original?

a. $\dfrac{3^x}{4}$ **b.** $\dfrac{3^x}{4^x}$ **c.** $\dfrac{3}{4^x}$

❷ To verify your answer to Step 1, use the table feature to compare the Y_1 and Y_2 columns. They should be exactly the same for every x-value.

❸ Based on Steps 1 and 2, to what part(s) of the quotient must the exponent be applied for the expression $\left(\dfrac{a}{b}\right)^x$?

Exercises

In Exerises 1–6, evaluate the expression.

1. $\left(\frac{4}{5}\right)^2$ **2.** $\left(\frac{5}{3}\right)^3$ **3.** $\left(-\frac{1}{2}\right)^4$

4. $\left(-\frac{5}{6}\right)^3$ **5.** $\left(\frac{7}{2}\right)^{-2}$ **6.** $\left(\frac{2}{3}\right)^{-4}$

See page 42 for keystrokes.

Lesson 8.3

NAME _____ DATE _____

Graphing Calculator Activity

For use with pages 463–469

TI-82

Y= (3 ÷ 4) ^ X,T,θ ENTER

3 ^ X,T,θ ÷ 4 ENTER

ZOOM 6

Y= ENTER CLEAR 3 ^ X,T,θ ÷ 4 ^

X,T,θ ENTER

GRAPH

Y= ENTER CLEAR 3 ÷ 4 ^ X,T,θ

ENTER

GRAPH

Enter the expression equivalent to $\left(\frac{3}{4}\right)^x$ in Y_2.

2nd [TblSet] (-) 3 ENTER 1 ENTER ENTER

▼ ENTER

2nd [TABLE]

TI-83

Y= (3 ÷ 4) ^ X,T,θ,n ENTER

3 ^ X,T,θ,n ÷ 4 ENTER

ZOOM 6

Y= ENTER CLEAR 3 ^ X,T,θ,n ÷ 4

^ X,T,θ,n ENTER

GRAPH

Y= ENTER CLEAR 3 ÷ 4 ^ X,T,θ,n

ENTER

GRAPH

Enter the expression equivalent to $\left(\frac{3}{4}\right)^x$ in Y_2.

2nd [TBLSET] (-) 3 ENTER 1 ENTER

ENTER ▼ ENTER

2nd [TABLE]

SHARP EL-9600c

Y= (3 ÷ 4) a^b X/θ/T/n ENTER

3 a^b X/θ/T/n ▶ ÷ 4 ENTER

ZOOM [A] 5

Y= ENTER CL 3 a^b X/θ/T/n ▶ ÷ 4

a^b X/θ/T/n ENTER

GRAPH

Y= ENTER CL 3 ÷ 4 a^b X/θ/T/n

ENTER

GRAPH

Enter the expression equivalent to $\left(\frac{3}{4}\right)^x$ in Y_2.

2ndF [TBLSET] ENTER ▼ 1 (-) 3

ENTER 1 ENTER TABLE

CASIO CFX-9850Ga PLUS

From the main menu, choose GRAPH.

(3 ÷ 4) ^ X,θ,T EXE

3 ^ X,θ,T ÷ 4 EXE

SHIFT F3 F3 EXIT F6

EXIT ▲ 3 ^ X,θ,T ÷ 4 ^ X,θ,T

EXE F6

EXIT ▲ 3 ÷ 4 ^ X,θ,T EXE F6

Enter the expression equivalent to $\left(\frac{3}{4}\right)^x$ in Y_2.

MENU 7 F5 (-) 3 EXE 3 EXE 1 EXE

EXIT F6

Algebra 1
Chapter 8 Resource Book

Lesson 8.3

NAME _____ DATE _____

Practice A

For use with pages 463–469

Use the quotient of powers property to simplify the expression.

1. $\dfrac{4^2}{4}$

2. $\dfrac{5^3}{5^5}$

3. $\dfrac{x^7}{x^3}$

4. $\dfrac{a^{10}}{a^5}$

5. $\dfrac{8^3}{8^5}$

6. $\dfrac{c^4}{c^6}$

7. $\dfrac{(-2)^3}{(-2)}$

8. $\dfrac{-(m^4)}{m^4}$

Use the power of a quotient property to simplify the expression.

9. $\left(\dfrac{1}{2}\right)^4$

10. $\left(\dfrac{2}{3}\right)^3$

11. $\left(\dfrac{4}{x}\right)^2$

12. $\left(\dfrac{3}{4}\right)^2$

13. $\left(\dfrac{3}{m}\right)^3$

14. $\left(\dfrac{x^2}{5}\right)^2$

15. $\left(\dfrac{3}{4}\right)^{-3}$

16. $\left(\dfrac{a^3}{b^2}\right)^4$

Evaluate the expression. Write your answer as a fraction in simplest form.

17. $\dfrac{7^5}{7^3}$

18. $\dfrac{6^5}{6^7}$

19. $\dfrac{18^6}{18^6}$

20. $\dfrac{(-5)^9}{5^9}$

21. $\dfrac{2^3}{2^{-4}}$

22. $\dfrac{4^5 \cdot 4^3}{4^6}$

23. $\left(\dfrac{2}{3}\right)^3$

24. $\left(\dfrac{3}{2}\right)^{-1}$

Simplify the expression. The simplified expression should have no negative exponents.

25. $\left(\dfrac{x}{3}\right)^4$

26. $\dfrac{x^7}{x^2}$

27. $\left(\dfrac{2}{x}\right)^6$

28. $x^5 \cdot \dfrac{1}{x^8}$

29. $x^{12} \cdot \dfrac{1}{x^3}$

30. $\left(\dfrac{x^5}{x^3}\right)^{-1}$

31. $\left(\dfrac{y^3}{y^5}\right)^{-2}$

32. $\dfrac{m^4 \cdot m^2}{m^7}$

33. $\dfrac{(t^3)^2}{(t^2)^3}$

34. $\dfrac{(2z)^4}{3z^2}$

35. $\dfrac{(2a^2b)^3}{(2ab^3)^2}$

36. $\left(\dfrac{3m^2n^4}{2mn^3}\right)^3$

Grade Point Average **In Exercises 37 and 38, use the following information.**

From Carmen's freshman year to her senior year, her grade point average (GPA) increased by approximately the same percentage each year. Carmen's GPA in year t can be modeled by

$$\text{GPA} = 2\left(\tfrac{6}{5}\right)^t, \text{ where } t = 0 \text{ corresponds to her freshman year.}$$

37. Complete the table showing Carmen's GPA throughout her high school career.

Year, t	0	1	2	3
GPA				

38. Find the ratio of Carmen's GPA in her senior year to her GPA in her sophomore year.

Lesson 8.3

NAME _____ DATE _____

Practice B

For use with pages 463–469

Use the quotient of powers property to simplify the expression.

1. $\dfrac{4^4}{4^2}$

2. $\dfrac{8^7}{8^9}$

3. $\dfrac{x^{15}}{x^9}$

4. $\dfrac{b^8}{b^{12}}$

5. $\dfrac{y^6}{y^0}$

6. $\dfrac{(-3)^7}{(-3)^3}$

7. $\dfrac{6^2 \cdot 6^{11}}{6^{16}}$

8. $\dfrac{x^{-8}}{x^{-5} \cdot x^{-4}}$

Use the power of a quotient property to simplify the expression.

9. $\left(\dfrac{1}{3}\right)^4$

10. $\left(\dfrac{5}{6}\right)^2$

11. $\left(\dfrac{4}{x}\right)^5$

12. $\left(\dfrac{y}{3}\right)^3$

13. $\left(\dfrac{7}{5}\right)^{-2}$

14. $\left(\dfrac{2^2}{a^5}\right)^3$

15. $\left(\dfrac{x^6}{y^3}\right)^8$

16. $\left(\dfrac{c^7}{d^{10}}\right)^4$

Evaluate the expression. Write your answer as a fraction in simplest form.

17. $\dfrac{2^8}{2^3}$

18. $\dfrac{-4^8}{(-4)^8}$

19. $\dfrac{5^{-2}}{5^{-5}}$

20. $\dfrac{7^{-2} \cdot 7^6}{(7^2)^2}$

21. $\dfrac{3^2 \cdot 3}{3^6}$

22. $\left(\dfrac{6}{7}\right)^{-2}$

23. $\left(\dfrac{12}{3}\right)^3$

24. $\left(-\dfrac{3}{8}\right)^2$

Simplify the expression. The simplified expression should have no negative exponents.

25. $\left(\dfrac{2}{x}\right)^5$

26. $\dfrac{1}{x^8} \cdot x^{20}$

27. $\left(\dfrac{b^{10}}{b^3}\right)^{-2}$

28. $\dfrac{r^{-5} \cdot r^5}{r^3}$

29. $\dfrac{(t^{-4})^9}{(t^{-4})^3}$

30. $\dfrac{(a^6 \cdot a^3)^3}{a^7}$

31. $\left(\dfrac{7x^{-2}y}{x^8 y^{-5}}\right)^3$

32. $\dfrac{-10xy^8}{2x^4 y^2} \cdot \dfrac{-5xy^{-2}}{(-y)^3}$

33. $\left(\dfrac{3x^7 y^9}{5x^5 y^2}\right)^{-4}$

34. $\dfrac{3xy^4}{2x^5 y} \cdot \dfrac{6x^{-3}y^2}{4y}$

35. $\dfrac{2x^2 y}{x^3 y^2} \cdot \dfrac{4x^7 y^2}{2x^3}$

36. $\left(\dfrac{4x^2 y^{-1}}{6xy}\right)^{-3} \cdot \dfrac{y^4}{x^6 y^2}$

Memory **In Exercises 37 and 38, use the following information.**

Suppose that you memorize a list of 100 German vocabulary words. Each week you forget $\frac{1}{8}$ of the words you knew the previous week. The number of vocabulary words V you remember after t weeks can be modeled by

$$V = 100\left(\tfrac{7}{8}\right)^t.$$

37. Complete the table showing the number of words you remember each week.

Week, t	0	5	10	15	20	25	30
Words, V							

38. Find the ratio of the number of words you remember in week 10 to the number of words you remember in week 25 without using the table.

NAME _____ DATE _____

Practice C

For use with pages 463–469

Use the quotient of powers property to simplify the expression.

1. $\dfrac{x^8}{x^{15}}$

2. $\dfrac{(-5)^{14}}{(-5)^{11}}$

3. $\dfrac{3^3 \cdot 3^9}{3^{17}}$

4. $\dfrac{x^{-7}}{x^5 \cdot x^{-14}}$

Use the power of a quotient property to simplify the expression.

5. $\left(\dfrac{3}{7}\right)^{-2}$

6. $\left(\dfrac{2^4}{x^5}\right)^{-1}$

7. $\left(\dfrac{x^4}{y^7}\right)^9$

8. $\left(\dfrac{a^{25}}{b^{14}}\right)^4$

Evaluate the expression. Write your answer as a fraction in simplest form.

9. $\dfrac{5^{16}}{5^{13}}$

10. $\dfrac{(-7)^{-3}}{-7^{-5}}$

11. $\dfrac{10^{-6}}{10^{-10}}$

12. $-\dfrac{9^{-4} \cdot 9^{-2}}{(9^{-3})^3}$

13. $\dfrac{4^4 \cdot 4^{-2}}{4^7}$

14. $\left(\dfrac{3^3}{5}\right)^{-2}$

15. $\left(\dfrac{36}{4}\right)^2$

16. $\left(-\dfrac{2}{4}\right)^5$

Simplify the expression. The simplified expression should have no negative exponents.

17. $\dfrac{6}{x^{10}} \cdot \dfrac{x^{17}}{15}$

18. $\left(\dfrac{y^{-5}}{y^9}\right)^{-4}$

19. $\dfrac{(a^{13} \cdot a^{-8})^5}{a^{31}}$

20. $\left(\dfrac{11x^6y^{-6}}{x^4y^{-3}}\right)^3$

21. $\left(\dfrac{2x^{-5}y^{12}}{3x^{-14}y^8}\right)^{-6}$

22. $\dfrac{(x^{-6})^{-3}}{(x^{-6})^2}$

23. $\dfrac{-12xy}{7x^4} \cdot \dfrac{21x^5y^2}{4y}$

24. $\dfrac{-3x^5}{x^{13}} \cdot \dfrac{2x^{10}y}{15y^2}$

25. $\dfrac{4xy^{11}}{x^7y^6} \cdot \dfrac{6x^8y}{8x^3}$

26. $\dfrac{y^{10}}{2x^3} \cdot \dfrac{20x^{14}}{xy^6}$

27. $\dfrac{5x^{-2}}{3x} \cdot \dfrac{2y^3}{x^{10}}$

28. $\left(\dfrac{5xy}{8x^{-1}y^2}\right)^2 \cdot \dfrac{26y^3}{5x^2y^5}$

29. $\dfrac{-8x^6y^{-3}}{3x^{-2}y^{-5}} \cdot \dfrac{-6x^{-10}y}{-4x}$

30. $\dfrac{4x^{-2}y^{-1}}{3x^{-3}} \cdot \dfrac{6x^{-3}y^{-2}}{8y^{-7}}$

31. $\left(\dfrac{2x^2y}{3y}\right) \cdot \left(\dfrac{4y^3}{x^4}\right)^2$

32. *Assembly Speed* An assembly-line worker increases the speed at which he can work by approximately the same percentage for the first 7 months of employment. The speed s (in parts assembled per hour) in t months can be modeled by $s = 10(1.01)^t$, where $t = 0$ corresponds to the month a worker was hired. Find the ratio of the speed of a worker after 7 months of experience to the speed of a worker after 4 months of experience.

33. *Population* The population P of California (in thousands) in 1995 projected through 2025 can be modeled by $P = 30{,}383(1.0157)^t$, where $t = 0$ represents 1995. Find the ratio of the population in 2025 to the population in 2000.

34. *Personal Computers* From 1982 to 1992, the cost of manufacturing a PC has decreased by about the same percentage each year. The cost C (in dollars) in year t can be modeled by $C = 3000\left(\frac{5}{6}\right)^t$, where $t = 0$ corresponds to 1992. Find the ratio of the cost in 1990 to the cost in 1985.

Reteaching with Practice

For use with pages 463–469

GOAL Use the division properties of exponents to evaluate powers and simplify expressions, and use the division properties of exponents to find a probability

---

VOCABULARY

Let a and b be numbers and let m and n be integers.

Quotient of Powers Property

To divide powers having the same base, subtract exponents.

$$\frac{a^m}{a^n} = a^{m-n}, \; a \neq 0 \qquad \text{Example: } \frac{3^7}{3^5} = 3^{7-5} = 3^2$$

Power of a Quotient Property

To find a power of a quotient, find the power of the numerator and the power of the denominator and divide.

$$\left(\frac{a}{b}\right)^m = \frac{a^m}{b^m}, \; b \neq 0 \qquad \text{Example: } \left(\frac{4}{5}\right)^3 = \frac{4^3}{5^3}$$

EXAMPLE 1 *Using the Quotient of Powers Property*

Use the quotient of powers property to simplify the expression.

a. $\dfrac{8^2 \cdot 8^4}{8^3}$

b. $z^7 \cdot \dfrac{1}{z^8}$

SOLUTION

To divide powers having the same base, subtract exponents.

a. $\dfrac{8^2 \cdot 8^4}{8^3} = \dfrac{8^6}{8^3}$

$\qquad = 8^{6-3}$

$\qquad = 8^3$

b. $z^7 \cdot \dfrac{1}{z^8} = \dfrac{z^7}{z^8}$

$\qquad = z^{7-8}$

$\qquad = z^{-1}$

$\qquad = \dfrac{1}{z}$

Exercises for Example 1

Use the quotient of powers property to simplify the expression.

1. $\dfrac{10^4}{10}$

2. $\dfrac{3^2}{3^3}$

3. $\dfrac{1}{y^2} \cdot y^8$

Algebra 1
Chapter 8 Resource Book

NAME _____ DATE _____

Reteaching with Practice

For use with pages 463–469

EXAMPLE 2 *Simplifying an Expression*

Simplify the expression. $\left(\dfrac{7a}{b^2}\right)^3$

SOLUTION

$$\left(\dfrac{7a}{b^2}\right)^3 = \dfrac{(7a)^3}{(b^2)^3} \qquad \text{Power of a quotient}$$

$$= \dfrac{7^3 \cdot a^3}{b^6} \qquad \text{Power of a product and power of a power}$$

$$= \dfrac{343a^3}{b^6} \qquad \text{Simplify.}$$

Exercises for Example 2

Simplify the expression. The simplified expression should have no negative exponents.

4. $\left(\dfrac{2}{x^3}\right)^4$
5. $\dfrac{z \cdot z^5}{z^2}$
6. $\left(\dfrac{5y^2}{w}\right)^2$

EXAMPLE 3 *Using the Power of a Quotient Property*

You toss a fair coin four times. Show that the probability of getting four tails is 0.0625.

SOLUTION

Probability that all four tosses are tails: $\left(\dfrac{1}{2}\right)^4$

Use the power of a quotient property to evaluate.

$$\left(\dfrac{1}{2}\right)^4 = \dfrac{1}{2^4} = \dfrac{1}{16} = 0.0625.$$

The probability of getting four tails is 0.0625.

Exercise for Example 3

7. You toss a fair coin six times. Show that the probability of getting six heads is about 0.0156.

NAME _____ DATE _____

Quick Catch-Up for Absent Students

For use with pages 463–469

The items checked below were covered in class on (date missed) _____

Lesson 8.3: Division Properties of Exponents

_____ **Goal 1:** Use the division properties of exponents to evaluate powers and to simplify expressions. (pp. 463–464)

Material Covered:

_____ Example 1: Using the Quotient of Powers Property

_____ Example 2: Using the Power of a Quotient Property

_____ Example 3: Simplifying Expressions

_____ **Goal 2:** Use the division property of exponents to solve real-life problems. (p. 465)

Material Covered:

_____ Example 4: Using the Quotient of Powers Property

_____ Student Help: Look Back

_____ Example 5: Using the Power of a Quotient Property

_____ Other (specify) _____

_____

Homework and Additional Learning Support

_____ Textbook (specify) _pp. 466–469_____

_____

_____ Internet: Extra Examples at www.mcdougallittel.com

_____ *Reteaching with Practice* worksheet (specify exercises)_____

_____ *Personal Student Tutor* for Lesson 8.3

NAME _____ DATE _____

Real-Life Application: When Will I Ever Use This?

For use with pages 463–469

Internet Usage

The Internet is a large network of computers that provides an enormous amount of information in the form of words, images, and sounds to users around the globe. Although the Internet was conceived in the early 1960s, it did not achieve its current form until the mid-1980s. Originally, the concept of the Internet was designed for military and research communication, however commercial use has made the Internet accessible to tens of millions of people in over 150 countries.

Today, the Internet provides numerous consumer services and information on almost any topic. News, weather, sports, discussion groups, banking, and shopping are just some of the features found on the Internet. Through the advent of electronic mail, or e-mail, messages can be sent from one computer to another with ease. As a result, telecommuting, or working from home, over the Internet allows people to live where they want, without the constraint of being close to work.

In Exercises 1–3, use the following information.

Consumer usage of the Internet is growing exponentially. One measure of this trend is the number of hours Americans spend on-line per person per year. The exponential equation for this growth is

$$y = 0.73 \cdot \left(\frac{8}{5}\right)^x,$$

where y is the number of hours per person per year and x is the number of years since 1990.

1. Find the number of hours consumers spend on-line per person per year in each of the following years. Round your answer to the nearest tenth.

   a. 1995

   b. 2001

2. Find the ratio of hours per person in 2001 to the hours per person in 1995.

3. Based on your answers to Exercises 1 and 2, do you think that the given exponential equation is a good model to use for prediction? Why or why not?

Challenge: Skills and Applications

For use with pages 463–469

In Exercises 1–3, find the value of n that makes the question true.

1. $\left(\dfrac{x^3}{x^5}\right)^n = x^{-6}$

2. $\dfrac{y^{4n-1}}{y^{2n}} = y^5$

3. $\left(\dfrac{y^{3n}}{y^{5n-4}}\right)^2 = y^{20}$

In Exercises 4–6, assuming the power of a power property $[(a^m)^n = a^{mn}]$ works for positive integers m and n, show that the property is true in each situation.

Example: m is a negative integer, n is a positive integer.

Solution: Let $m = -k$, where k is a positive integer. Then

$$(a^m)^n = (a^{-k})^n = \left(\dfrac{1}{a^k}\right)^n = \dfrac{1^n}{(a^k)^n} = \dfrac{1}{a^{kn}} = a^{-kn} = a^{mn}$$

4. m is a positive integer, n is a negative integer.

5. m is a negative integer, n is a negative integer.

6. m is a positive integer, n is zero.

In Exercises 7–9, use the following information.

The 17th-century French mathematician Pierre de Fermat proved a theorem about any prime number p and any integer a that is not divisible by p. You can discover this fact for $p = 5$ and $a = 2, 3, 4, 6,$ and 7 by calculating the values of $2^4, 3^4, 4^4, 6^4,$ and 7^4.

7. What remainders do you get when you divide your answers by 5?

8. If $p = 7$, you can verify Fermat's theorem by calculating $2^6, 3^6, 4^6,$ $5^6,$ and 6^6. What remainders do you get when you divide these numbers by 7?

9. Based on your answers from Exercises 7 and 8, state Fermat's theorem. Start your statement with, "If p is a prime number and a is not divisible by p..."

NAME _____ DATE _____

Quiz 1

For use after Lessons 8.1–8.3

1. Simplify $(2x^2y)^3(-3xy^2)^2$. *(Lesson 8.1)*

2. Which is larger, 4^3 or 3^4? *(Lesson 8.1)*

3. Solve the equation for x. *(Lesson 8.1)*

$$(4^2)^5 = 4^x$$

4. Evaluate the expression $6^{-7} \cdot 6^{10}$. *(Lesson 8.2)*

5. Rewrite the expression using positive exponents. *(Lesson 8.2)*

$$(-5y)^{-3}$$

6. Evaluate $(4y)(7^0)$. *(Lesson 8.2)*

7. Evaluate the expression $\left(-\frac{2}{5}\right)^3$. *(Lesson 8.3)*

8. Simplify $\dfrac{2xy^4}{5xy^2} \cdot \dfrac{-30xy}{2x^2y}$. *(Lesson 8.3)*

Answers

1. _____

2. _____

3. _____

4. _____

5. _____

6. _____

7. _____

8. _____

Lesson 8.3

TEACHER'S NAME _____ CLASS _____ ROOM _____ DATE _____

Lesson Plan

1-day lesson (See *Pacing the Chapter,* TE pages 446C–446D)   For use with pages 470–475

GOALS 1. Use scientific notation to represent numbers.
2. Use scientific notation to describe real-life situations.

State/Local Objectives _____

_____

✓ Check the items you wish to use for this lesson.

STARTING OPTIONS
____ Homework Check: TE page 466; Answer Transparencies
____ Warm-Up or Daily Homework Quiz: TE pages 470 and 469, CRB page 54, or Transparencies

TEACHING OPTIONS
____ Lesson Opener (Application): CRB page 55 or Transparencies
____ Graphing Calculator Activity with Keystrokes: CRB page 56
____ Examples 1–6: SE pages 470–472
____ Extra Examples: TE pages 471–472 or Transparencies; Internet
____ Closure Question: TE page 472
____ Guided Practice Exercises: SE page 473

APPLY/HOMEWORK
Homework Assignment
____ Basic   16–56 even, 64, 65, 70, 75, 78, 80, 82
____ Average   16–56 even, 61, 64, 65, 70, 75, 78, 80, 82
____ Advanced   16–56 even, 60–66, 70, 75, 78, 80, 82

Reteaching the Lesson
____ Practice Masters: CRB pages 57–59 (Level A, Level B, Level C)
____ Reteaching with Practice: CRB pages 60–61 or Practice Workbook with Examples
____ Personal Student Tutor

Extending the Lesson
____ Cooperative Learning Activity: CRB page 63
____ Applications (Interdisciplinary): CRB page 64
____ Challenge: SE page 475; CRB page 65 or Internet

ASSESSMENT OPTIONS
____ Checkpoint Exercises: TE pages 471–472 or Transparencies
____ Daily Homework Quiz (8.4): TE page 475, CRB page 68, or Transparencies
____ Standardized Test Practice: SE page 475; TE page 475; STP Workbook; Transparencies

Notes _____

_____

_____

LESSON 8.4

Lesson Plan for Block Scheduling

Half-day lesson (See *Pacing the Chapter*, TE pages 446C–446D)   For use with pages 470–475

GOALS
1. Use scientific notation to represent numbers.
2. Use scientific notation to describe real-life situations.

State/Local Objectives _____

_____

_____

✓ **Check the items you wish to use for this lesson.**

STARTING OPTIONS

____ Homework Check: TE page 466; Answer Transparencies

____ Warm-Up or Daily Homework Quiz: TE pages 470 and
          469, CRB page 54, or Transparencies

CHAPTER PACING GUIDE	
Day	**Lesson**
1	Assess Ch. 7; 8.1 (begin)
2	8.1 (end); 8.2 (begin)
3	8.2 (end); 8.3 (begin)
4	8.3 (end); **8.4 (all)**
5	8.5 (all); 8.6 (begin)
6	8.6 (end); Review Ch. 8
7	Assess Ch. 8; 9.1 (all)

TEACHING OPTIONS

____ Lesson Opener (Application): CRB page 55 or Transparencies

____ Graphing Calculator Activity with Keystrokes: CRB page 56

____ Examples 1–6: SE pages 470–472

____ Extra Examples: TE pages 471–472 or Transparencies; Internet

____ Closure Question: TE page 472

____ Guided Practice Exercises: SE page 473

APPLY/HOMEWORK

Homework Assignment (See also the assignment for Lesson 8.3.)

____ Block Schedule: 16–56 even, 61, 64, 65, 70, 75, 78, 80, 82

Reteaching the Lesson

____ Practice Masters: CRB pages 57–59 (Level A, Level B, Level C)

____ Reteaching with Practice: CRB pages 60–61 or Practice Workbook with Examples

____ Personal Student Tutor

Extending the Lesson

____ Cooperative Learning Activity: CRB page 63

____ Applications (Interdisciplinary): CRB page 64

____ Challenge: SE page 475; CRB page 65 or Internet

ASSESSMENT OPTIONS

____ Checkpoint Exercises: TE pages 471–472 or Transparencies

____ Daily Homework Quiz (8.4): TE page 475, CRB page 68, or Transparencies

____ Standardized Test Practice: SE page 475; TE page 475; STP Workbook; Transparencies

Notes _____

_____

_____

Lesson 8.4

WARM-UP EXERCISES

For use before Lesson 8.4, pages 470–475

Evaluate each expression.

1. 10^3

2. 10^{-4}

3. 100^2

4. 10^{-8}

5. 10^{10}

··

DAILY HOMEWORK QUIZ

For use after Lesson 8.3, pages 463–469

Evaluate the expression. Write fractions in simplest form.

1. $\dfrac{7^5}{7^3}$ 2. $\left(-\dfrac{3}{5}\right)^4$ 3. $\left(\dfrac{9}{4}\right)^{-1}$

Simplify the expression. The simplified expression should have no negative exponents.

4. $b^7 \cdot \dfrac{b^4}{b^3}$ 5. $\dfrac{-9m^4n}{27mn^2}$ 6. $\dfrac{-5x^{-3}y^4}{x^2y^{-1}} \cdot \dfrac{(3x^2y^2)^3}{x^3y}$

7. The amount of money after t years in an account that begins with $700 and that earns compound interest at a rate of 4.7% per year can be modeled by the equation $A = 700(1.047)^t$. Find the ratio of the amount after 6 years to the amount after 4 years. Express the ratio as a power of 1.047.

Algebra 1
Chapter 8 Resource Book

Application Lesson Opener

For use with pages 470–475

Raul made Table 1 on the left. Rita made Table 2. Rita says that even though the numbers in their tables look different, they have the same value. Is Rita correct? Prove your answer.

Table 1

Planet	Average distance from sun (in miles)
Mercury	36,000,000
Venus	67,200,000
Earth	92,750,000
Mars	141,300,000
Jupiter	483,600,000
Saturn	887,000,000
Uranus	1,780,000,000
Neptune	2,794,000,000
Pluto	3,658,000,000

Table 2

Planet	Average distance from sun (in miles)
Mercury	3.6×10^7
Venus	6.72×10^7
Earth	9.275×10^7
Mars	1.413×10^8
Jupiter	4.836×10^8
Saturn	8.87×10^8
Uranus	1.78×10^9
Neptune	2.794×10^9
Pluto	3.658×10^9

Graphing Calculator Activity

For use with page 471

Keystrokes for Example 4

TI-82

7.48 [2nd] [EE] [(-)] 7
[×] 2.4 [2nd] [EE] 9 [ENTER]

TI-83

7.48 [2nd] [EE] [(-)] 7
[×] 2.4 [2nd] [EE] 9 [ENTER]

SHARP EL-9600c

7.48 [2ndF] [Exp] [(-)] 7 [×]
2.4 [2ndF] [Exp] 9 [ENTER]

CASIO CFX-9850Ga PLUS

From the main menu, choose RUN.

7.48 [SHIFT] [10^x] [(-)] 7 [×]
2.4 [SHIFT] [10^x] 9 [EXE]

Lesson 8.4

Practice A

For use with pages 470–475

Write each power of a ten as a numeral.

1. 10^3
2. 10^5
3. 10^{-2}
4. 10^2
5. 10^{-3}
6. 10^{-1}
7. 10^4
8. 10^{-5}

Rewrite in decimal form.

9. 2.03×10^3
10. 3.4578×10^4
11. 6.43×10^1
12. 7.2×10^5
13. 5.2×10^0
14. 4.68×10^{-2}
15. 1.3×10^{-6}
16. 8.497×10^{-3}
17. 9.8×10^{-4}

Rewrite in scientific notation.

18. 25,000
19. 36.41
20. 4,000,000
21. 564,200
22. 9.32
23. 0.15
24. 0.0083
25. 0.000000718
26. 0.0673

Evaluate the expression without a calculator. Write the result in decimal form.

27. $2 \times 10^3 \cdot 3 \times 10^8$
28. $3 \times 10^{-4} \cdot 3 \times 10^{-5}$
29. $2 \times 10^{-5} \cdot 3 \times 10^7$
30. $4 \times 10^{-6} \cdot 2 \times 10^5$
31. $3 \times 10^6 \cdot 4 \times 10^3$
32. $7 \times 10^{-3} \cdot 5 \times 10^{-1}$
33. $3 \times 10^5 \cdot 8 \times 10^{-2}$
34. $12 \times 10^3 \cdot 3 \times 10^{-6}$
35. $6 \times 10^{-8} \cdot 7 \times 10^6$

Write the number in scientific notation.

36. *Earth to Pluto*   As the planets orbit the sun, the closest Pluto gets to Earth is approximately 2,700,000,000 miles.

37. *Speed of Light*   The speed of light in a vacuum is approximately 186,000 miles per second.

38. *Pluto's Diameter*   The diameter of Pluto is approximately 1400 miles. There are 5280 feet in one mile. What is the diameter of Pluto in feet?

39. *Fingernails*   Human fingernails grow at a rate of about 0.00286 inch per day.

40. *Television sets*   As of 1996, the United States has the most homes with television sets with 95,400,000.

41. *Cells*   The human body has 1×10^{12} cells. There are 3×10^{10} red blood cells. Find the ratio of red blood cells to the total number of cells.

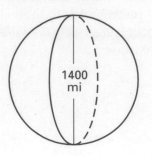

1400 mi

Practice B

For use with pages 470–475

Rewrite in decimal form.

1. 3.79×10^5

2. 2.5×10^{-2}

3. 8.44×10^1

4. 6.5393×10^4

5. 3.589×10^{-3}

6. 9.1187×10^0

7. 1.0056×10^{-5}

8. 7.2658746×10^8

9. $4.33652157 \times 10^{11}$

Rewrite in scientific notation.

10. 64.3

11. 0.0602

12. 998.653

13. $22,078,600$

14. 45.668

15. $63,000,000$

16. 0.007485

17. 0.0000056388

18. $7,960,000,000$

Evaluate the expression without a calculator. Write the result in decimal form.

19. $(5 \times 10^{-3}) \cdot (3 \times 10^6)$

20. $(9 \times 10^{-4}) \cdot (4 \times 10^{-1})$

21. $(6.2 \times 10^{-5}) \cdot (4.4 \times 10^7)$

22. $\dfrac{5 \times 10^{-2}}{8 \times 10^{-6}}$

23. $\dfrac{2.16 \times 10^3}{3.6 \times 10^{-7}}$

24. $\dfrac{7.7 \times 10^{-4}}{1.1 \times 10^{-1}}$

25. $(4 \times 10^{-4})^{-1}$

26. $(5 \times 10^{-2})^3$

27. $(2.1 \times 10^3)^2$

Write the number in scientific notation.

28. *Neptune to Sun* The mean distance of Neptune to the sun is 2,796,460,000 miles.

29. *Red Blood Cells* The thickness of a red blood cell is approximately 0.0003125 inch.

30. *Shoes* Americans buy more than 850,000,000 pairs of shoes each year.

31. *Largest Country* The largest country is Russia. It has a total area of 6,592,800 square miles.

32. *Pixels* The images on a computer screen are made up of more than 5000 pixels, or dots, per square inch. How many pixels are on a computer screen that is 154 square inches?

33. *Smallest Newspaper* The world's smallest newspaper is a Bengali newspaper called "Bireswar-Smriti." It has an area of 0.00756 square foot.

34. *Blinking* Consider a person who blinks about 20,000 times per day and who lives to be 76 years old. Estimate the number of times the person blinks during his or her lifetime. Do not acknowledge leap years. Write your answer in decimal form and in scientific notation.

35. *California* California has an area of approximately 1.56×10^5 square miles. California has a population of about 2.98×10^7. How many people are there per square mile?

NAME _____ DATE _____

Practice C

For use with pages 470–475

Rewrite in decimal form.

1. 9.99×10^6

2. 1.356×10^{-3}

3. 6.01×10^1

4. 7.00563×10^4

5. 3.2×10^{-4}

6. 3.50086×10^0

7. $7.78895233146 \times 10^{10}$

8. 0.206230×10^{-6}

9. 8.623514×10^{13}

Rewrite in scientific notation.

10. 0.032

11. $33,254.236$

12. $87,000$

13. 0.0000222

14. 0.00099578

15. $70,000,000$

16. 0.852246

17. 0.000000008

18. $12,840,000,000$

Evaluate the expression without a calculator. Write the result in decimal form.

19. $(9 \times 10^{-3}) \cdot (8 \times 10^{-6})$

20. $(4 \times 10^7) \cdot (1.2 \times 10^{-2})$

21. $(6.8 \times 10^2) \cdot (5 \times 10^3)$

22. $(4.2 \times 10^{-6}) \cdot (2 \times 10^5)$

23. $(3.1 \times 10^6) \cdot (4 \times 10^3)$

24. $(0.7 \times 10^{-5}) \cdot (5.4 \times 10^{-1})$

25. $\dfrac{3 \times 10^5}{8 \times 10^{-2}}$

26. $\dfrac{12.6 \times 10^3}{3 \times 10^{-2}}$

27. $\dfrac{4.96 \times 10^{-7}}{6.2 \times 10^{-4}}$

28. $(4.0 \times 10^3)^2$

29. $(2 \times 10^{-3})^3$

30. $(1.25 \times 10^{-1})^{-4}$

Write the number in scientific notation.

31. **Human Cells**   The body of a human has more than 1,000,000,000,000 cells.

32. **Human Ears**   The human ear grows about 0.0087 inch in one year. How much does the ear grow in one month?

33. **Earth's Diameter**   The polar diameter of Earth is approximately 7900 miles. There are approximately 161,000 cm in one mile. What is the polar diameter of Earth in cm?

34. **Mass of Helium Atom**   A proton (P) and a neutron (N) each weigh 1.67×10^{-24} gram. An electron weighs 9.11×10^{-28} gram. Find the mass of one helium atom.

35. **Surface Area**   The total surface area of Earth is about 1.97×10^8 square miles. The surface area of land on Earth is about 5.73×10^7 square miles. Find the ratio of surface area of land to that of the entire planet.

NAME _____ DATE _____

Reteaching with Practice

 GOAL Use scientific notation to represent numbers and to describe real-life situations

---

VOCABULARY

A number is written in **scientific notation** if it is of the form $c \times 10^n$, where $1 \le c < 10$ and n is an integer.

EXAMPLE 1 *Rewriting in Decimal Form*

Rewrite each number in decimal form.

a. 2.23×10^4 **b.** 8.5×10^{-3}

SOLUTION

a. $2.23 \times 10^4 = 22,300$ Move decimal point right 4 places.

b. $8.5 \times 10^{-3} = 0.0085$ Move decimal point left 3 places.

Exercises for Example 1

Rewrite each number in decimal form.

1. 9.332×10^6 **2.** 2.78×10^{-1} **3.** 4.5×10^5

EXAMPLE 2 *Rewriting in Scientific Notation*

Rewrite each number in scientific notation.

a. 0.0729 **b.** $26,645$

SOLUTION

a. $0.0729 = 7.29 \times 10^{-2}$ Move decimal point right 2 places.

b. $26,645 = 2.6645 \times 10^4$ Move decimal point left 4 places.

Exercises for Example 2

Rewrite each number in scientific notation.

4. 75.2 **5.** $135,667$ **6.** 0.00088

Lesson 8.4

NAME _____ DATE _____

Reteaching with Practice

For use with pages 470–475

EXAMPLE 3 *Computing with Scientific Notation*

Evaluate the expression and write the result in scientific notation.

$$(7.0 \times 10^4)^2$$

SOLUTION

To multiply, divide, or find powers of numbers in scientific notation, use the properties of exponents.

$(7.0 \times 10^4)^2 = 7.0^2 \times (10^4)^2$ Power of a product

$\qquad\qquad\quad = 49 \times 10^8$ Power of a power

$\qquad\qquad\quad = 4.9 \times 10^9$ Write in scientific notation.

Exercises for Example 3

Evaluate the expression and write the result in scientific notation.

7. $(2.3 \times 10^{-1})(5.5 \times 10^3)$ **8.** $(2.0 \times 10^{-1})^3$

EXAMPLE 4 *Dividing with Scientific Notation*

The mass of the sun is approximately 1.99×10^{30} kilograms. The mass of the moon is approximately 7.36×10^{22} kilograms. The mass of the sun is approximately how many times that of the moon?

SOLUTION

Find the ratio of the mass of the sun to the mass of the moon.

$$\frac{1.99 \times 10^{30}}{7.36 \times 10^{22}} \approx 0.27 \times 10^8$$

$$\doteq 2.7 \times 10^7$$

The mass of the sun is about 27,000,000 times that of the moon.

Exercise for Example 4

9. The Pacific Ocean covers about 1.66241×10^8 square kilometers. The Baltic Sea covers about 4.144×10^5 square kilometers. The Pacific Ocean is approximately how many times as large as the Baltic Sea?

NAME _____ DATE _____

Quick Catch-Up for Absent Students

For use with pages 470–475

The items checked below were covered in class on (date missed) _____

Lesson 8.4: Scientific Notation

____ **Goal 1:** Use scientific notation to represent numbers. (pp. 470–471)

Material Covered:

____ Activity: Investigating Scientific Notation

____ Example 1: Rewriting in Decimal Form

____ Example 2: Rewriting in Scientific Notation

____ Example 3: Computing with Scientific Notation

____ Student Help: Keystroke Help

____ Example 4: Using a Calculator

Vocabulary:

scientific notation, p. 470

____ **Goal 2:** Use scientific notation to describe real-life situations. (p. 472)

Material Covered:

____ Example 5: Multiplying with Scientific Notation

____ Student Help: Look Back

____ Example 6: Dividing with Scientific Notation

____ Other (specify) _____

_____

Homework and Additional Learning Support

____ Textbook (specify) pp. 473–475 _____

_____

____ Internet: Extra Examples at www.mcdougallittel.com

____ *Reteaching with Practice* worksheet (specify exercises)_____

____ *Personal Student Tutor* for Lesson 8.4

Lesson 8.4

Cooperative Learning Activity

For use with pages 470–475

GOAL To use scientific notation to write the distances of the planets from the sun

Materials: paper, pencil

Exploring Scientific Notation

In this activity, you and your partner will research the distances of the nine planets in our solar system from the sun. You will use scientific notation to write these distances.

Instructions

❶ Research the distances of all nine planets in our solar system from the sun. Write down each distance in decimal form and scientific notation.

❷ Make a diagram that shows these distances. The diagram should be drawn to scale.

❸ Present your findings to the class.

Analyzing the Results

1. How much farther is Pluto from the sun than Mercury? Write your answer in scientific notation.

2. How many times farther is Pluto from the sun than Mercury? Write your answer in scientific notation.

3. How many times Earth's distance from the sun is Jupiter? Write your answer in scientific notation.

NAME _____ DATE _____

Interdisciplinary Application

For use with pages 470–475

Sahara Desert

GEOGRAPHY   The Sahara, a region in northern Africa, is the world's largest desert. It extends about 16,896,000 feet from the Atlantic Ocean to the Red Sea and about 5,808,000 feet from the Atlas Mountains and the Mediterranean Sea to the Niger River Valley, an area roughly equal to that of the United States. In addition to spreading over all of the Western Sahara and Egypt, the Sahara includes parts of Morocco, Algeria, Tunisia, Libya, Sudan, Chad, Niger, Mali, and Mauritania.

Although most people think of the desert as a desolate expanse of sand, sand covers only about 25% of the Sahara. Plateaus of rock, areas of gravel, mountains, and oases make up the majority of the landscape. The 2 million people who inhabit the Sahara are mainly nomads with herds of sheep and goats. They use camels to travel between the major oases for water, supplies, and trade.

In Exercises 1-5, use the information above.

1. Rewrite the length and width of the Sahara in scientific notation.

2. Use your answers from Exercise 1 to find the approximate area of the Sahara in square feet.

3. Find the area of the Sahara that is covered with sand.

4. Assume that the average depth of the sand is 200 feet. The volume of the sand is equal to the area of sand times its depth. Find the volume of the sand in cubic feet.

5. A grain of sand has a volume of about 1.3×10^{-9} cubic foot. Use your answer from Exercise 4 to estimate how many grains of sand are in the Sahara Desert.

Lesson 8.4

NAME _____ DATE _____

Challenge: Skills and Applications

For use with pages 470–475

In Exercises 1–3, evaluate the expression without using a calculator. Write the result in decimal form.

1. $(3.55 \times 10^8) + (1.6 \times 10^6)$

2. $7.94 \times 10^{-7} - \dfrac{1.2 \times 10^{-6}}{4}$

3. $4(6.25 \times 10^7) + 1.4 \times 10^8$

In Exercises 4–5, write the answer in scientific notation.

4. Find a number 100,000 times as large as 3.95×10^4.

5. Find a number 100,000 times as large as 3.95×10^n.

6. Which of the following two numbers is larger? How many times as large is it?

    2.5×10^{-5} 1.25×10^2

In Exercises 7–8, use the following information.

The radius of Earth is about 6.38×10^3 kilometers. The radius of the sun is about 6.96×10^5 kilometers. The planet Jupiter has a radius of about 7.14×10^4 kilometers.

7. The sun's volume is about how many times as large as Earth's volume? (The formula for the volume of a sphere is $V = \dfrac{4}{3}\pi r^3$.)

8. Jupiter's volume is about how many times as large as Earth's volume?

TEACHER'S NAME _____ CLASS _____ ROOM _____ DATE _____

Lesson Plan

1-day lesson (See *Pacing the Chapter,* TE pages 446C–446D)   **For use with pages 476–482**

GOALS   1. **Write and use models for exponential growth.**
2. **Graph models for exponential growth.**

State/Local Objectives _____

_____

✓ Check the items you wish to use for this lesson.

STARTING OPTIONS
____ Homework Check: TE page 473; Answer Transparencies
____ Warm-Up or Daily Homework Quiz: TE pages 477 and 475, CRB page 68, or Transparencies

TEACHING OPTIONS
____ Motivating the Lesson: TE page 478
____ Concept Activity: SE page 476; CRB page 69 (Activity Support Master)
____ Lesson Opener (Application): CRB page 70 or Transparencies
____ Graphing Calculator Activity with Keystrokes: CRB pages 71–72
____ Examples 1–5: SE pages 477–479
____ Extra Examples: TE pages 478–479 or Transparencies
____ Closure Question: TE page 479
____ Guided Practice Exercises: SE page 480

APPLY/HOMEWORK
Homework Assignment
____ Basic   6–16, 18–23, 29, 30, 35, 40, 44, 45
____ Average   6–16, 18–24, 29, 30, 35, 40, 44, 45
____ Advanced   6–16, 18–31, 35, 40, 44, 45

Reteaching the Lesson
____ Practice Masters: CRB pages 73–75 (Level A, Level B, Level C)
____ Reteaching with Practice: CRB pages 76–77 or Practice Workbook with Examples
____ Personal Student Tutor

Extending the Lesson
____ Applications (Real-Life): CRB page 79
____ Challenge: SE page 482; CRB page 80 or Internet

ASSESSMENT OPTIONS
____ Checkpoint Exercises: TE pages 478–479 or Transparencies
____ Daily Homework Quiz (8.5): TE page 482, CRB page 83, or Transparencies
____ Standardized Test Practice: SE page 482; TE page 482; STP Workbook; Transparencies

Notes _____

_____

_____

TEACHER'S NAME _____ CLASS _____ ROOM _____ DATE _____

Lesson Plan for Block Scheduling

Half-day lesson (See *Pacing the Chapter*, TE pages 446C–446D)   For use with pages 476–482

GOALS 1. **Write and use models for exponential growth.**
2. **Graph models for exponential growth.**

State/Local Objectives _____

_____

_____

✓ **Check the items you wish to use for this lesson.**

STARTING OPTIONS

____ Homework Check: TE page 473; Answer Transparencies
____ Warm-Up or Daily Homework Quiz: TE pages 477 and
          475, CRB page 68, or Transparencies

CHAPTER PACING GUIDE	
Day	**Lesson**
1	Assess Ch. 7; 8.1 (begin)
2	8.1 (end); 8.2 (begin)
3	8.2 (end); 8.3 (begin)
4	8.3 (end); 8.4 (all)
5	**8.5 (all)**; 8.6 (begin)
6	8.6 (end); Review Ch. 8
7	Assess Ch. 8; 9.1 (all)

TEACHING OPTIONS

____ Motivating the Lesson: TE page 478
____ Concept Activity: SE page 476; CRB page 69 (Activity Support Master)
____ Lesson Opener (Application): CRB page 70 or Transparencies
____ Graphing Calculator Activity with Keystrokes: CRB pages 71–72
____ Examples 1–5: SE pages 477–479
____ Extra Examples: TE pages 478–479 or Transparencies
____ Closure Question: TE page 479
____ Guided Practice Exercises: SE page 480

APPLY/HOMEWORK

Homework Assignment (See also the assignment for Lesson 8.6.)

____ Block Schedule: 6–16, 18–24, 29, 30, 35, 40, 44, 45

Reteaching the Lesson

____ Practice Masters: CRB pages 73–75 (Level A, Level B, Level C)
____ Reteaching with Practice: CRB pages 76–77 or Practice Workbook with Examples
____ Personal Student Tutor

Extending the Lesson

____ Applications (Real-Life): CRB page 79
____ Challenge: SE page 482; CRB page 80 or Internet

ASSESSMENT OPTIONS

____ Checkpoint Exercises: TE pages 478–479 or Transparencies
____ Daily Homework Quiz (8.5): TE page 482, CRB page 83, or Transparencies
____ Standardized Test Practice: SE page 482; TE page 482; STP Workbook; Transparencies

Notes _____

_____

_____

NAME _____ DATE _____

WARM-UP EXERCISES

For use before Lesson 8.5, pages 476–482

Evaluate when $x = 5$.

1. $2 + 3.1x$

2. $2 + (3.1)^x$

3. $-12 + 9x$

4. $-12 + 9^x$

DAILY HOMEWORK QUIZ

For use after Lesson 8.4, pages 470–475

Rewrite in decimal form.

1. 6.32×10^{-5} **2.** 4.55×10^0 **3.** 7.168×10^7

Rewrite in scientific notation.

4. 3591.76

5. 0.0000849

6. 460,000,000

**Evaluate the expression without using a calculator.
Write the result in decimal form.**

7. $(9 \times 10^{-12}) \cdot (6 \times 10^3)$

8. $\dfrac{1.8 \times 10^{-2}}{2.4 \times 10^{-7}}$

Algebra 1
Chapter 8 Resource Book

LESSON 8.5

Activity Support Master

For use with page 476

Step 1

x	0	1	2	3	4	5
y	20	25				

Step 2

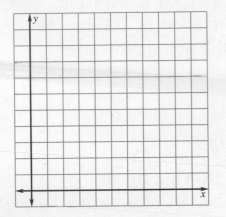

Step 3

x	0	1	2	3	4	5
y	1	5				

Step 4

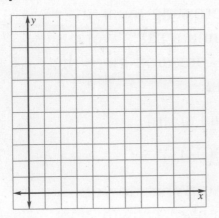

Step 5

_____

_____

_____

In Questions 1–3, use a calculator and the given information to predict which expression will give you the correct value for the growth over time.

1. The population of a small town has been growing by an average of 2% a year for the past 5 years. That means an increase of 2% the first year, then the new population increased by 2% for the second year, and so on. The town had a population of 700 before the growth began. Which expression gives the population after the 5 years of growth? Explain.

   A. $700(0.02)(5)$

   B. $700(1.02)$

   C. $700(1.02)(5)$

   D. $700(1.02)^5$

2. The amount earned by a business has been growing by an average of 5% a year for the past 3 years. The business earned $50,000 before the growth began. Which expression gives the amount earned by the business at the end of the 3-year period? Explain.

   A. $50,000(1.05)$

   B. $50,000(1.05)^3$

   C. $50,000(1.05)(3)$

   D. $50,000(0.05)^3$

3. The sales by a catalog company have been growing by an average of 3% a year for the past 4 years. The annual sales were $100,000 before the growth began. Which expression gives the sales at the end of the 4-year period? Explain.

   A. $100,000(1.03)^4$

   B. $100,000(1.03)(4)$

   C. $100,000(0.03)(4)$

   D. $100,000(0.03)^4$

4. Compare the answers to Questions 1–3 above. What do you notice about the beginning amount, the percent of growth, and the time period in each answer?

LESSON 8.5

NAME _____ DATE _____

Graphing Calculator Activity

For use with pages 477–482

GOAL **To develop the Rule of 72**

In Lesson 8.5 you are going to learn about exponential growth functions. The variable in these functions represents the time period. One of the most common applications of exponential growth functions is compound interest.

Activity

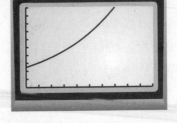

❶ You deposit $100 in a savings account that pays 2% annual interest compounded yearly. The function that models your account balance over time is $y = 100(1.02)^x$. Enter this function into your graphing calculator and plot the graph.

❷ You want to find out how long it will take your money to double. Use the *Trace* or *Table* features to find the value of x when $y = \$200$.

❸ Repeat Steps 1 and 2 for each of the annual interest rates below.

a. 3% annual interest
$y = 100(1.03)^x$

b. 6% annual interest
$y = 100(1.06)^x$

c. 9% annual interest
$y = 100(1.09)^x$

Use the interest rates from Steps 1 and 3. Find the quotient of 72 and each interest rate. (Do not use the decimal form of the percent.) What do you notice?

Exercises

1. Rule of 72: If a quantity is growing at $r\%$ per year, then the quantity will _____ in approximately $72 \div r$ years.

In Exercises 2–5, use the Rule of 72 to find how long it would take to double an initial deposit at the given interest rate.

2. 4%   **3.** 8%   **4.** 12%   **5.** 10%

See page 72 for keystrokes.

Algebra 1
Chapter 8 Resource Book

71

LESSON 8.5 CONTINUED

Graphing Calculator Activity

For use with pages 477–482

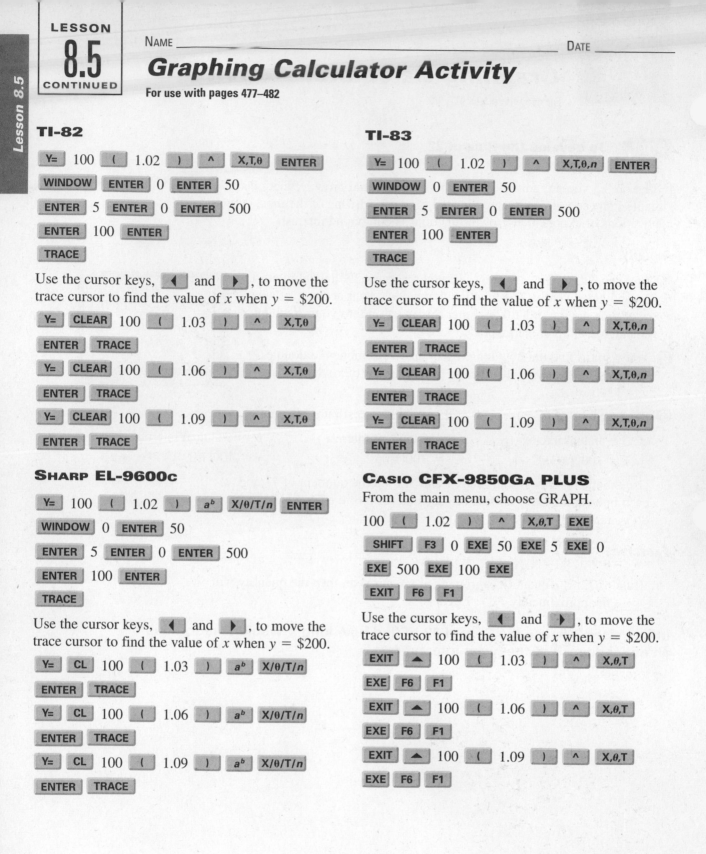

TI-82

Y= 100 (1.02) ^ X,T,θ ENTER

WINDOW ENTER 0 ENTER 50

ENTER 5 ENTER 0 ENTER 500

ENTER 100 ENTER

TRACE

Use the cursor keys, ◀ and ▶, to move the trace cursor to find the value of x when $y = \$200$.

Y= CLEAR 100 (1.03) ^ X,T,θ

ENTER TRACE

Y= CLEAR 100 (1.06) ^ X,T,θ

ENTER TRACE

Y= CLEAR 100 (1.09) ^ X,T,θ

ENTER TRACE

TI-83

Y= 100 (1.02) ^ X,T,θ,n ENTER

WINDOW 0 ENTER 50

ENTER 5 ENTER 0 ENTER 500

ENTER 100 ENTER

TRACE

Use the cursor keys, ◀ and ▶, to move the trace cursor to find the value of x when $y = \$200$.

Y= CLEAR 100 (1.03) ^ X,T,θ,n

ENTER TRACE

Y= CLEAR 100 (1.06) ^ X,T,θ,n

ENTER TRACE

Y= CLEAR 100 (1.09) ^ X,T,θ,n

ENTER TRACE

SHARP EL-9600c

Y= 100 (1.02) a^b X/θ/T/n ENTER

WINDOW 0 ENTER 50

ENTER 5 ENTER 0 ENTER 500

ENTER 100 ENTER

TRACE

Use the cursor keys, ◀ and ▶, to move the trace cursor to find the value of x when $y = \$200$.

Y= CL 100 (1.03) a^b X/θ/T/n

ENTER TRACE

Y= CL 100 (1.06) a^b X/θ/T/n

ENTER TRACE

Y= CL 100 (1.09) a^b X/θ/T/n

ENTER TRACE

CASIO CFX-9850GA PLUS

From the main menu, choose GRAPH.

100 (1.02) ^ X,θ,T EXE

SHIFT F3 0 EXE 50 EXE 5 EXE 0

EXE 500 EXE 100 EXE

EXIT F6 F1

Use the cursor keys, ◀ and ▶, to move the trace cursor to find the value of x when $y = \$200$.

EXIT ▲ 100 (1.03) ^ X,θ,T

EXE F6 F1

EXIT ▲ 100 (1.06) ^ X,θ,T

EXE F6 F1

EXIT ▲ 100 (1.09) ^ X,θ,T

EXE F6 F1

NAME _____ DATE _____

Practice A

For use with pages 477–482

Write the growth factor used to model each percent of increase in an exponential function.

1. 3%

2. 9%

3. 6.8%

4. 10.2%

5. 1.01%

6. 0.8%

You deposit $1300 in an account that pays 4% interest compounded yearly. Find the balance for the given time period.

7. 3 years

8. 6 years

9. 10 years

10. 25 years

Find the balance after 5 years of an account that pays 5.2% interest compounded yearly given the following investment amounts.

11. $500

12. $650

13. $1000

14. $1250

15. *Money Choices* Which option gives the greater ending balance?

  a. Put $100 in an account that pays 5% interest compounded yearly for 5 years.

  b. Keep $105 in your room and add $4 to it each year for 5 years.

16. *Money Choices* Which option gives the greater ending balance?

  a. Put $625 in an account that pays 8% interest compounded yearly for 8 years.

  b. Put $750 in an account that pays 6% interest compounded yearly for 8 years.

17. *College Tuition* From 1990 to 2000, the tuition at a college increased by about 8% per year. Use the graph below to write an exponential growth equation.

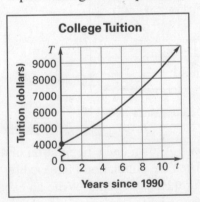

College Tuition

Profit Increase **In Exercises 18 and 19, use the following information.**

A business had an $11,000 profit in 1995. Then the profit increased by 15% per year for the next 5 years.

18. Write an exponential growth model.

19. Graph the exponential growth model in Exercise 18.

Write an exponential growth model.

20. A business had a $5000 profit in 1990. Then the profit increased by 15% per year for the next 10 years.

21. A town had a population of 48,000 in 1995. Then the population increased by 8% per year for the next 5 years.

22. You memorized 15 Spanish vocabulary words. Then you memorized 20% more words each week for the next 10 weeks.

NAME _____

DATE _____

Practice B

For use with pages 477–482

You deposit $1500 in an account that pays 5% interest compounded yearly. Find the balance for the given time period.

1. 3 years   **2.** 5 years   **3.** 8 years   **4.** 10 years

Find the balance after 5 years of an account that pays 5.2% interest compounded yearly given the following investment amounts.

5. $500   **6.** $550   **7.** $600   **8.** $800

9. *Money Choices* Which option gives the greater ending balance?

  a. Put $360 in an account that pays 5% interest compounded yearly for 10 years.

  b. Put $375 in an account that pays 4.5% interest compounded yearly for 10 years.

10. *Money Choices* Which option gives the greater ending balance?

  a. Put $100 in an account that pays 8% interest compounded yearly for 10 years.

  b. Put $150 in an account that pays 7% interest compounded yearly for 5 years.

11. *Profit Increases* From 1990 to 2000, the profit earned by a company increased by about 1.5% per year. Use the graph below to write an exponential growth equation.

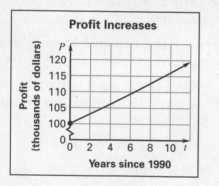

Profit Increases

12. *California* The population of California increased about 32% per decade from 1930 to 1990. Use the graph below to write an exponential growth model. If the population continues to increase, what will the population be in 2000?

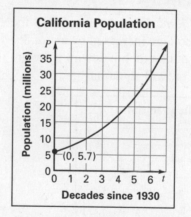

California Population

13. *Business Profit* A business had a $5000 profit in 1990. Then the profit increased by 15% per year for the next 10 years. Write an exponential growth model for the company's profit over the 10-year period. Use the model to find the profit in 1995.

14. *Population* A town had a population of 48,000 in 1995. Then the population increased by 8% per year for the next 5 years. Write an exponential growth model for the town's population over the 5-year period. Use the model to find the population in 1998.

15. *Memorizing Vocabulary* You memorized 15 Spanish vocabulary words. Then you memorized 20% more words each week for the next 10 weeks. Write an exponential growth model for the number of vocabulary words you memorized over the 10-week period. Use the model to find the number of vocabulary words you knew after 6 weeks.

16. *Fruit Fly Population* You begin with 50 fruit flies for a science project. The fruit flies increase in number by 30% each day for a week. Write an exponential growth model for the number of fruit flies over the one-week period. Use the model to find the number of fruit flies you have at the end of 4 days.

Practice C

For use with pages 477–482

You deposit $1250 in an account that pays 3.9% interest compounded yearly. Find the balance for the given time period.

   1. 5 years   **2.** 10 years   **3.** 15 years   **4.** 20 years

Find the balance after 7 years of an account that pays 6.15% interest compounded yearly given the following investment amounts.

   5. $900   **6.** $1550   **7.** $3000   **8.** $4500

9. *Investment* What is the value of an $10,000 investment after 1 year if it earns 7% annual interest compounded *quarterly*? To solve, use the compound interest formula

$$P = 10,000\left(1 + \frac{0.07}{n}\right)^n$$

where n is the total number of compounding periods.

10. *Profit Increase* From 1990 to 2000, the profit earned by a company increased by about 2.25% per year. Use the graph below to write an exponential growth equation.

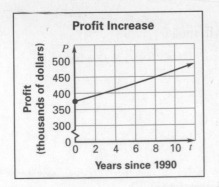

Profit Increase

Profit (thousands of dollars)

Years since 1990

11. Copy and complete each table of values. Sketch the graph of the exponential function.

  a. $A = 2000(1.08)^t$

t	1	2	3	4	5	6
A						

  b. $y = 3500(1.055)^x$

x	1	2	3	4	5	6
y						

12. *Business Profit* A business had a $5000 profit in 1990. Then the profit increased by 12.5% per year for the next 10 years. Write an exponential growth model for the company's profit over the 10-year period. Use the model to find the profit in 1995.

13. *Population* A town had a population of 48,000 in 1995. Then the population increased by 8.3% per year for the next 5 years. Write an exponential growth model for the town's population over the 5-year period. Use the model to find the population in 1998.

14. *Fruit Flies* A population of 25 fruit flies is released in the school's science lab. The population doubles each day for 5 days. What is the percent of increase each day? What is the population after 5 days?

15. *Internet Users* The number of students who have applied for Internet privileges at school has tripled each month. What is the percent of increase each month? If 45 students had Internet privileges initially, how many had privileges 4 months later?

NAME _____ DATE _____

Reteaching with Practice

For use with pages 477–482

GOAL Write and use models for exponential growth and graph models for exponential growth

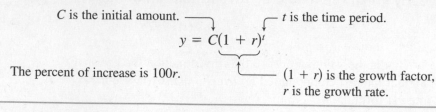

VOCABULARY

Exponential growth occurs when a quantity increases by the same percent in each unit of time.

C is the initial amount. ⟶   ⟵ t is the time period.

$$y = C(1 + r)^t$$

The percent of increase is $100r$. $(1 + r)$ is the growth factor, r is the growth rate.

EXAMPLE 1 *Finding the Balance in an Account*

A principal of $600 is deposited in an account that pays 3.5% interest compounded yearly. Find the account balance after 4 years.

SOLUTION

Use the exponential growth model to find the account balance A.
The growth rate is 0.035. The initial value is 600.

$A = P(1 + r)^t$	Exponential growth model
$= 600(1 + 0.035)^4$	Substitute 600 for P, 0.035 for r, and 4 for t.
$= 600(1.035)^4$	Simplify.
≈ 688.514	Evaluate.

The balance after 4 years will be about $688.51.

Exercises for Example 1

Use the exponential growth model to find the account balance.

1. A principal of $450 is deposited in an account that pays 2.5% interest compounded yearly. Find the account balance after 2 years.

2. A principal of $800 is deposited in an account that pays 3% interest compounded yearly. Find the account balance after 5 years.

NAME _____ DATE _____

Reteaching with Practice

For use with pages 477–482

EXAMPLE 2 *Writing an Exponential Growth Model*

A population of 40 pheasants is released in a wildlife preserve.
The population doubles each year for 3 years. What is the population
after 4 years?

SOLUTION

Because the population doubles each year, the growth factor is 2.
Then $1 + r = 2$, and the growth rate $r = 1$.

$$P = C(1 + r)^t \qquad \text{Exponential growth model}$$
$$= 40(1 + 1)^4 \qquad \text{Substitute for } C, r, \text{ and } t.$$
$$= 40 \cdot 2^4 \qquad \text{Simplify.}$$
$$= 640 \qquad \text{Evaluate.}$$

After 4 years, the population will be about 640 pheasants.

Exercise for Example 2

3. A population of 50 pheasants is released in a wildlife preserve.
The population triples each year for 3 years. What is the population
after 3 years?

EXAMPLE 3 *Graphing an Exponential Growth Model*

Graph the exponential growth model in Example 2.

SOLUTION

Make a table of values, plot the points in a coordinate
plane, and draw a smooth curve through the points.

t	0	1	2	3	4	5
P	40	80	160	320	640	1280

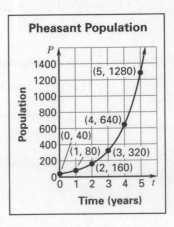

Pheasant Population

Exercise for Example 3

4. Graph the exponential growth model in Exercise 3.

NAME _____ DATE _____

Quick Catch-Up for Absent Students

For use with pages 476–482

The items checked below were covered in class on (date missed) _____

Activity 8.5: Linear and Exponential Growth Models (p. 476)

_____ **Goal:** Determine how linear growth models and exponential growth models are different.

Lesson 8.5: Exponential Growth Functions

_____ **Goal 1:** Write and use models for exponential growth. (pp. 477–478)

Material Covered:

_____ Example 1: Finding the Balance in an Account

_____ Example 2: Writing an Exponential Growth Model

_____ Student Help: Study Tip

_____ Example 3: Writing an Exponential Growth Model

Vocabulary:

exponential growth, p. 477   growth factor, p. 477
initial amount, p. 477   time period, p. 477
percent of increase, p. 477   growth rate, p. 477

_____ **Goal 2:** Graph models for exponential growth. (p. 479)

Material Covered:

_____ Example 4: A Model with a Small Growth Factor

_____ Example 5: A Model with a Large Growth Factor

_____ Other (specify) _____

_____

Homework and Additional Learning Support

_____ Textbook (specify) pp. 480–482 _____

_____

_____ *Reteaching with Practice* worksheet (specify exercises) _____

_____ *Personal Student Tutor* for Lesson 8.5

Real-Life Application: When Will I Ever Use This?

For use with pages 477–482

Investing for College

Saving enough money for college can seem impossible, especially when many sources claim it will cost $120,000 or more for four years. Though many Ivy League institutions are this expensive, other public and private schools can be as low as $13,000 for four years. Even this lower amount requires financial planning. Fortunately, there are many ways to invest money that can be used for college in the future. The best savings plans are those that are started when a child is born. If this is not possible, the sooner you can begin to save, the more money you will have available to you when you actually begin your journey into higher education.

Roth IRAs, traditionally individual retirement accounts, can be used for qualified higher-education expenses after an account has been open for five successive tax years. Owners may then take tax-free and penalty-free distributions of earnings. Educational IRAs can be established as well. Like a Roth IRA, parents or grandparents can make annual contributions, which are not tax deductible. The money, however, is allowed to grow tax-free until it is withdrawn for higher education. The drawback is that there is presently a limit of only $500 a year per student.

Traditional savings accounts and certificates can also be opened with banks or credit unions.

1. Imagine that your parents established a savings account for you when you were born. They deposited $1500 with an interest rate of 3.54% compounded yearly. How much balance will you have to further your education at age 18?

2. Certificates of deposit (CDs), which are guaranteed for a specific amount of time, generally offer a higher percentage rate than savings accounts. When you entered sixth grade, your parents opened a $2000 seven-year CD with 5.87% interest rate that is compounded yearly. How much would your balance be after high school graduation?

3. Imagine that you could have the same percentage rate in your savings account that you had in your certificate of deposit. How much would your balance be at age 18? (Use your original $1500 deposit from Exercise 1.)

4. Graph the exponential growth model in Exercise 1. Let the horizontal axis represent the number of years, with $t = 0$ being your birth year. Let the vertical axis represent the amount of money in the balance.

5. Graph the exponential growth model in Exercise 2. Let the horizontal axis represent the number of years, with $t = 0$ being your birth year. Let the vertical axis represent the amount of money in the balance. (Make a table of values using your exponential growth model from Exercise 2 for seven years.)

Challenge: Skills and Applications

For use with pages 477–482

In Exercises 1–3, use the following information.

In the 1990 census, the population of San Diego, California, was 1,110,623. In the 1980 census it was 875,538.

1. Assuming exponential growth, write an equation for the population y of San Diego t decades after 1980. Round the growth rate r to the nearest thousandth.

2. Use the equation from Exercise 1 to estimate the population of San Diego in 2010.

3. Use the equation from Exercise 1 to estimate the population of San Diego in 1950.

In Exercises 4–8, you will write and apply a function for the growth of bacteria.

4. Write a function that models bacteria growth y over t time periods for growth rate r and initial population C.

5. Use the function from Exercise 4 to find the population when $C = 10,000$, $r = 20\%$, and $t = 3.8$.

6. Use the function from Exercise 4 to find the population when $C = 25,000$, $r = 14\%$, and $t = 6.7$.

7. Use a graph and the function from Exercise 4 to estimate how many time periods it takes for the bacteria population to reach 21,000 for $C = 7000$ and $r = 8\%$.

8. Use a graph and the function from Exercise 4 to estimate how many time periods it takes for the bacteria population to reach 32,000 for $C = 16,000$ and $r = 18\%$.

In Exercises 9–11, use the following information to find the frequency of each note. Use the full calculator display until you round to the nearest tenth for the final answer.

The frequencies of the notes of the musical scale grow exponentially as the notes increase by half-steps. The note middle A has a frequency of 440 Hertz (Hz). The note B-flat, one half-step above middle A, has a frequency of about 466.2 Hertz.

9. the C 3 half-steps above middle A

10. the E 7 half-steps above middle A

11. the A 12 half-steps above middle A

TEACHER'S NAME _____ CLASS _____ ROOM _____ DATE _____

Lesson Plan

2-day lesson (See *Pacing the Chapter,* **TE pages 446C–446D)**   **For use with pages 483–492**

GOALS 1. Write and use models for exponential decay.
2. Graph models for exponential decay.

State/Local Objectives _____

_____

✓ **Check the items you wish to use for this lesson.**

STARTING OPTIONS

____ Homework Check: TE page 480; Answer Transparencies
____ Warm-Up or Daily Homework Quiz: TE pages 484 and 482, CRB page 83, or Transparencies

TEACHING OPTIONS

____ Concept Activity: SE page 483; CRB page 84 (Activity Support Master)
____ Lesson Opener (Application): CRB page 85 or Transparencies
____ Graphing Calculator Activity with Keystrokes: CRB pages 86–87
____ Examples: Day 1: 1–3, SE pages 484–485; Day 2: 4–6, SE pages 486–487
____ Extra Examples: Day 1: TE page 485 or Transp.; Day 2: TE pages 486–487 or Transp.
____ Technology Activity: SE page 492
____ Closure Question: TE page 487
____ Guided Practice: SE page 488; Day 1: Exs. 1–8; Day 2: Ex. 9

APPLY/HOMEWORK
Homework Assignment

____ Basic   Day 1: 10–25; Day 2: 26–33, 35, 36, 41–43, 45, 48; Quiz 2: 1–10
____ Average   Day 1: 10–25; Day 2: 26–33, 35, 36, 41–43, 45, 48; Quiz 2: 1–10
____ Advanced   Day 1: 10–25; Day 2: 26–43, 45, 48; Quiz 2: 1–10

Reteaching the Lesson

____ Practice Masters: CRB pages 88–90 (Level A, Level B, Level C)
____ Reteaching with Practice: CRB pages 91–92 or Practice Workbook with Examples
____ Personal Student Tutor

Extending the Lesson

____ Cooperative Learning Activity: CRB page 94
____ Applications (Real-Life): CRB page 95
____ Math & History: SE page 491; CRB page 96; Internet
____ Challenge: SE page 490; CRB page 97 or Internet

ASSESSMENT OPTIONS

____ Checkpoint Exercises: Day 1: TE page 485 or Transp.; Day 2: TE pages 486–487 or Transp.
____ Daily Homework Quiz (8.6): TE page 490 or Transparencies
____ Standardized Test Practice: SE page 490; TE page 490; STP Workbook; Transparencies
____ Quizzes (8.4–8.6): SE page 491; CRB page 98

Notes _____

_____

_____

LESSON

8.6

Lesson Plan for Block Scheduling

1-day lesson (See *Pacing the Chapter*, TE pages 446C–446D)   For use with pages 483–492

  1. Write and use models for exponential decay.
2. Graph models for exponential decay.

State/Local Objectives _____

_____

_____

✓ **Check the items you wish to use for this lesson.**

STARTING OPTIONS

____ Homework Check: TE page 480; Answer Transparencies
____ Warm-Up or Daily Homework Quiz: TE pages 484 and
         482, CRB page 83, or Transparencies

CHAPTER PACING GUIDE	
Day	**Lesson**
1	Assess Ch. 7; 8.1 (begin)
2	8.1 (end); 8.2 (begin)
3	8.2 (end); 8.3 (begin)
4	8.3 (end); 8.4 (all)
5	8.5 (all); **8.6 (begin)**
6	**8.6 (end)**; Review Ch. 8
7	Assess Ch. 8; 9.1 (all)

TEACHING OPTIONS

____ Concept Activity: SE page 483; CRB page 84 (Activity Support Master)
____ Lesson Opener (Application): CRB page 85 or Transparencies
____ Graphing Calculator Activity with Keystrokes: CRB pages 86–87
____ Examples: Day 5: 1–3, SE pages 484–485; Day 6: 4–6, SE pages 486–487
____ Extra Examples: Day 5: TE page 485 or Transp.; Day 6: TE pages 486–487 or Transp.
____ Technology Activity: SE page 492
____ Closure Question: TE page 487
____ Guided Practice: SE page 488; Day 5: Exs. 1–8; Day 6: Ex. 9

APPLY/HOMEWORK

Homework Assignment (See also the assignment for Lesson 8.5.)

____ Block Schedule: Day 5: 10–25; Day 6: 26–33, 35, 36, 41–43, 45, 48; Quiz 2: 1–10

Reteaching the Lesson

____ Practice Masters: CRB pages 88–90 (Level A, Level B, Level C)
____ Reteaching with Practice: CRB pages 91–92 or Practice Workbook with Examples
____ Personal Student Tutor

Extending the Lesson

____ Cooperative Learning Activity: CRB page 94
____ Applications (Real-Life): CRB page 95
____ Math & History: SE page 491; CRB page 96; Internet
____ Challenge: SE page 490; CRB page 97 or Internet

ASSESSMENT OPTIONS

____ Checkpoint Exercises: Day 5: TE page 485 or Transp.; Day 6: TE pages 486–487 or Transp.
____ Daily Homework Quiz (8.6): TE page 490 or Transparencies
____ Standardized Test Practice: SE page 490; TE page 490; STP Workbook; Transparencies
____ Quizzes (8.4–8.6): SE page 491; CRB page 98

Notes _____

_____

_____

82

Algebra 1
Chapter 8 Resource Book

WARM-UP EXERCISES

For use before Lesson 8.6, pages 483–492

Evaluate each expression. Round to the nearest hundredth.

1. $(0.85)^5$

2. $(0.97)^4$

3. $(0.5)^3$

4. $(1 - 0.03)^4$

5. $(1 - 0.08)^6$

....

DAILY HOMEWORK QUIZ

For use after Lesson 8.5, pages 476–482

1. You deposit $700 in an account that pays 8% interest compounded yearly. Find the balance after 13 years.

2. How much must you deposit in an account that pays 8% interest compounded yearly to have a balance of $2000 after 6 years?

3. Company A had a $15,000 profit. Then the profit increased by 20% per year for the next 15 years. Company B had a $25,000 profit. Then the profit increased by 16% per year for the next 15 years. Write an exponential growth model for each situation.

Activity Support Master

For use with page 483

Lesson 8.6

Number of toss	0	1	2	3	4	5	6	7	8	9	10	11
Number of pennies remaining												

NAME _____ DATE _____

Application Lesson Opener

For use with pages 484–491

The table shows the decrease in the number of VCRs sold each year by an electronics store chain.

Year	Number of VCRs sold
1	1,000,000
2	950,000
3	902,500
4	857,375

1. Has the number of VCRs sold decreased by the same amount each year? Explain how you know.

2. Has the number of VCRs sold decreased by the same percent each year? Explain how you know.

3. If this pattern continues, how many VCRs would you expect the chain to sell in year 5? year 6?

The table shows the decrease in the number of large packages delivered by a delivery service over the past 3 years.

Year	Number of large packages delivered
1	1000
2	800
3	640

4. Has either the number or the percent decreased by the same amount each year? Explain how you know.

5. If this pattern continues, how many large packages would you expect the service to deliver in year 4? in year 5?

Graphing Calculator Activity Keystrokes

For use with page 490

Keystrokes for Exercise 37

TI-82

Y= 4 ^ X,T,θ ENTER

0.25 ^ X,T,θ ENTER

2nd [TblSet] (-) 3 ENTER 1 ENTER ENTER

▼ ENTER 2nd [TABLE]

ZOOM 6

TI-83

Y= 4 ^ X,T,θ,n ENTER

0.25 ^ X,T,θ,n ENTER

2nd [TblSet] (-) 3 ENTER 1 ENTER ENTER

▼ ENTER 2nd [TABLE]

ZOOM 6

SHARP EL-9600c

Y= 4 a^b X/θ/T/n ENTER

0.25 a^b X/θ/T/n ENTER

2ndF [TBLSET] ENTER ▼

(-) 3 ENTER 1 ENTER TABLE

ZOOM [A] 5

CASIO CFX-9850GA PLUS

From the main menu, choose TABLE.

F5 (-) 3 EXE 3 EXE 1 EXE EXIT

4 ^ X,θ,T EXE

0.25 ^ X,θ,T EXE F6

MENU 5

SHIFT F3 F3 EXIT

F6

Algebra 1
Chapter 8 Resource Book

NAME _____ DATE _____

Graphing Calculator Activity Keystrokes

For use with Technology Activity 8.6 on page 492

TI-82

STAT 1

Enter *x*-values in L1.

0 ENTER 1 ENTER 2 ENTER 3 ENTER 4 ENTER
5 ENTER 6 ENTER 7 ENTER 8 ENTER 9 ENTER

Enter *x*-values in L2.

0.82 ENTER 0.64 ENTER 0.5 ENTER
0.39 ENTER 0.3 ENTER 0.24 ENTER
0.18 ENTER 0.14 ENTER 0.11 ENTER
0.08 ENTER

STAT ▶ ALPHA [A] 2nd [L1] , 2nd [L2]
ENTER
Y= VARS 5 ▶ ▶ 7
WINDOW ENTER 0 ENTER 20
ENTER 1 ENTER 0 ENTER 1 ENTER 0.1 ENTER
GRAPH

TI-83

STAT 1

Enter *x*-values in L1.

0 ENTER 1 ENTER 2 ENTER 3 ENTER 4 ENTER
5 ENTER 6 ENTER 7 ENTER 8 ENTER 9 ENTER

Enter *x*-values in L2.

0.82 ENTER 0.64 ENTER 0.5 ENTER
0.39 ENTER 0.3 ENTER 0.24 ENTER
0.18 ENTER 0.14 ENTER 0.11 ENTER
0.08 ENTER

STAT ▶ 0 2nd [L1] , 2nd [L2]
ENTER
Y= VARS 5 ▶ ▶ 1
WINDOW 0 ENTER 20 ENTER 1 ENTER
0 ENTER 1 ENTER 0.1 ENTER
GRAPH

SHARP EL-9600c

STAT [A] ENTER

Enter *x*-values in L1.

0 ENTER 1 ENTER 2 ENTER 3 ENTER 4 ENTER
5 ENTER 6 ENTER 7 ENTER 8 ENTER 9 ENTER

Enter *x*-values in L2.

0.82 ENTER 0.64 ENTER 0.5 ENTER 0.39 ENTER
0.3 ENTER 0.24 ENTER 0.18 ENTER 0.14 ENTER
0.11 ENTER 0.08 ENTER

2ndF [QUIT]

STAT [D] 0 9

(2ndF [L1] , 2ndF [L2] , VARS [A]
ENTER [A] 1) ENTER
WINDOW 0 ENTER 20 ENTER 1 ENTER
0 ENTER 1 ENTER 0.1 ENTER
GRAPH

CASIO CFX-9850GA PLUS

From the main menu, choose STAT.

Enter *x*-values in List 1.

0 EXE 1 EXE 2 EXE 3 EXE 4 EXE
5 EXE 6 EXE 7 EXE 8 EXE 9 EXE

Enter *x*-values in List 2.

0.82 EXE 0.64 EXE 0.5 EXE 0.39 EXE
0.3 EXE 0.24 EXE 0.18 EXE 0.14 EXE
0.11 EXE 0.08 EXE

SHIFT F3 0 EXE 20 EXE 1 EXE 0 EXE 1
EXE 0.1 EXE EXIT
F1 F6

Choose the following:
  Graph Type: Scatter, XList: List1; YList: List 2;
Frequency: 1; Mark Type: ▫

EXIT

F1 F6 F2 F6

Algebra 1
Chapter 8 Resource Book

Practice A

For use with pages 484–491

Match the graph with its equation.

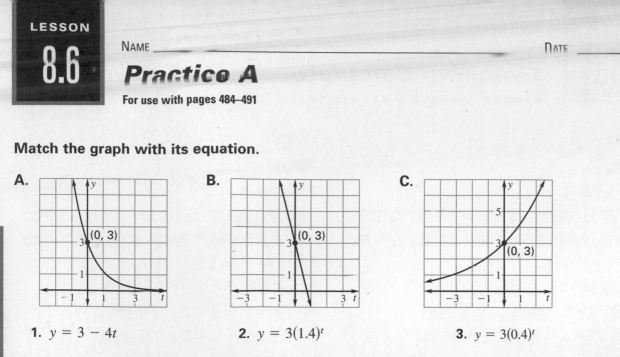

A. (0, 3)

B. (0, 3)

C. (0, 3)

1. $y = 3 - 4t$

2. $y = 3(1.4)^t$

3. $y = 3(0.4)^t$

Car Value **You buy a used car for $10,000. It depreciates at the rate of 20% per year. Find the value of the car for the given year.**

4. 1 year

5. 3 years

6. 5 years

7. 8 years

Recognizing Models **Classify the model as exponential growth or exponential decay. Identify the growth or decay factor and the percent increase or decrease per time period.**

8. $y = 3(1.7)^t$

9. $y = 10(0.2)^t$

10. $y = 2\left(\frac{1}{2}\right)^t$

11. $y = 12(2)^t$

12. $y = 16\left(\frac{7}{6}\right)^t$

13. $y = 100(1.05)^t$

14. *Population*   Between 1970 and 2000, the population of a city decreased by approximately 2% each year. In 1970 there were 600,000 people. What was the population in 2000?

15. *Losses*   Between 1990 and 2000, the profits of a business decreased by approximately 0.5% each year. In 1990 the business's profit was $2 million. Write an exponential decay model showing the business's profit P in year t. What was the profit in 2000?

16. *Declining Employment*   TRL Industries had 3000 employees in 1990. Each year for 10 years, the number of employees decreased by 3%. Complete the table showing the number of employees for each year.

17. Sketch a graph of the results of Exercise 16.

Year	1990 ($t = 0$)	1991 ($t = 1$)	1992 ($t = 2$)	1993 ($t = 3$)	1994 ($t = 4$)	1995 ($t = 5$)
Number of Employees						

Year	1996 ($t = 6$)	1997 ($t = 7$)	1998 ($t = 8$)	1999 ($t = 9$)	2000 ($t = 10$)
Number of Employees					

18. If the number of employees at TRL Industries continues to decrease at the same rate, how many employees does the model predict in 2005?

NAME _____ DATE _____

Practice B

For use with pages 484–491

Car Value **You buy a used car for $15,000. It depreciates at the rate of 23% per year. Find the value of the car for the given year.**

1. 1 year **2.** 3 years **3.** 5 years **4.** 8 years

Classify the model as exponential growth or exponential decay. Identify the growth or decay factor and the percent increase or decrease per time period.

5. $y = 16(1.25)^t$ **6.** $y = 800\left(\frac{6}{5}\right)^t$ **7.** $y = 4(0.94)^t$

8. $y = 32(0.75)^t$ **9.** $y = 10\left(\frac{4}{5}\right)^t$ **10.** $y = 20(1.05)^t$

11. *Population* Between 1985 and 2000, the population of a city decreased by approximately 3% each year. In 1985 there were 450,000 people. What was the population in 2000?

12. *Losses* Between 1990 and 2000, the profits of a business decreased by approximately 0.7% each year. In 1990 the business's profit was $1.4 million. Write an exponential decay model showing the business's profit P in year t. What was the profit in 2000?

13. *Unemployment Rate* In 1998 the unemployment rate of a city decreased by approximately 1.2% each month. In January the unemployment rate was 8%. What was the rate in December?

Copy and complete the table of values and sketch a graph of the exponential function.

14. $y = 2(0.25)^x$

x	-3	-2	-1	0	1	2
y						

15. $y = 0.5(0.6)^x$

x	-3	-2	-1	0	1	2
y						

Unemployment Rate

Unemployment rate (percent)

Months since January

16. *Declining Employment* TRL Industries had 14,000 employees in 1990. Each year for 10 years, the number of employees decreased by 4%. Complete the table showing the number of employees for each year.

17. Sketch a graph of the results of Exercise 16.

Year	1990 ($t = 0$)	1991 ($t = 1$)	1992 ($t = 2$)	1993 ($t = 3$)	1994 ($t = 4$)	1995 ($t = 5$)
Number of Employees						

Year	1996 ($t = 6$)	1997 ($t = 7$)	1998 ($t = 8$)	1999 ($t = 9$)	2000 ($t = 10$)
Number of Employees					

18. If the number of employees at TRL Industries continues to decrease at the same rate, how many employees does the model predict in 2005?

NAME _____ DATE _____

Practice C

For use with pages 484–491

Truck Value **You buy a used truck for $17,000. It depreciates at the rate of 19% per year. Find the value of the truck for the given year.**

1. 1 year   **2.** 3 years   **3.** 5 years   **4.** 8 years

Recognizing Models **Classify the model as exponential growth or exponential decay. Identify the growth or decay factor and the percent increase or decrease per time period.**

5. $y = 0.05(10)^t$

6. $y = \frac{1}{3}(3)^t$

7. $y = 29(0.75)^t$

8. $y = 156(1.01)^t$

9. $y = 1.5\left(\frac{2}{3}\right)^t$

10. $y = 296\left(\frac{9}{5}\right)^t$

11. *Population*   Between 1970 and 2000, the population of a city decreased by approximately 4% each year. In 1970 there was a population of 870,000. What was the population in 2000?

12. *Sleeping Behavior*   On an average, as people grow older, they sleep fewer hours during the night. The amount of sleep that your grandfather gets has decreased by 1.5% each year since 1980. Using the graph at the right, write an exponential decay model showing the number of hours your grandfather sleeps per night. How many hours per night did he sleep in 2000?

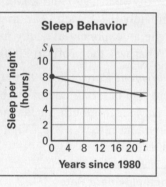

13. *Health Club*   A health club had a declining enrollment from 1995 to 2000. The enrollment in 1995 was 1285 people. Each year for the next five years, the enrollment decreased by 2.3%. Make a table showing the enrollment for each year. Sketch a graph of the results.

14. *Declining Employment*   TRL Industries had 14,000 employees in 1990. Each year for 10 years, the number of employees decreased by 3.5%. Complete the table showing the number of employees for each year. Sketch a graph of the results.

Year	1990	1991	1992	1993	1994	1995	1996	1997	1998	1999	2000
Number of Employees											

15. If the number of employees at TRL Industries continues to decrease at the same rate, how many employees does the model predict in 2005?

16. *Challenge*   You invested $1000 in 1990. It increased 5% each year for five years. Over the next five years it decreased in value by 5% each year. Will you have $1000 again at the end of ten years? Explain your reasoning.

Reteaching with Practice

For use with pages 484–491

GOAL Write and use models for exponential decay and graph models for exponential decay

VOCABULARY

Exponential decay occurs when a quantity decreases by the same percent in each unit of time.

C is the initial amount. $\longrightarrow$ $\qquad$ t is the time period.

$$y = C(1 - r)^t$$

The percent of decrease is $100r$. $\qquad$ $(1 - r)$ is the decay factor, r is the decay rate.

EXAMPLE 1 *Writing an Exponential Decay Model*

You bought a used truck for $15,000. The value of the truck will decrease each year because of depreciation. The truck depreciates at the rate of 8% per year. Write an exponential decay model to represent the real-life problem.

SOLUTION

The initial value C is $15,000. The decay rate r is 0.08. Let y be the value and let t be the number of years you have owned the truck.

$$y = C(1 - r)^t \qquad \text{Exponential decay model}$$
$$= 15,000(1 - 0.08)^t \qquad \text{Substitute 15,000 for } C \text{ and 0.08 for } r.$$
$$= 15,000(0.92)^t \qquad \text{Simplify.}$$

The exponential decay model is $y = 15,000(0.92)^t$.

Exercises for Example 1

1. Use the exponential decay model in Example 1 to estimate the value of your truck in 5 years.

2. Use the exponential decay model in Example 1 to estimate the value of your truck in 7 years.

3. Rework Example 1 if the truck depreciates at the rate of 10% per year.

NAME _____ DATE _____

Reteaching with Practice

For use with pages 484–491

EXAMPLE 2 *Graphing an Exponential Decay Model*

   a. Graph the exponential decay model in Example 1.

   b. Use the graph to estimate the value of your truck in 6 years.

SOLUTION

   a. Make a table of values to verify the model in Example 1. Find
      the value of the truck for each year by multiplying the value in
      the previous year by the decay factor $1 - 0.08 = 0.92$.

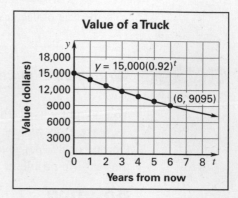

Year	Value
0	15,000
1	$0.92(15{,}000) = 13{,}800$
2	$0.92(13{,}800) = 12{,}696$
3	$0.92(12{,}696) \approx 11{,}680$
4	$0.92(11{,}680) \approx 10{,}746$
5	$0.92(10{,}746) \approx 9886$

Use the table of values to write ordered pairs:
(0, 15,000), (1, 13,800), (2, 12,696), (3, 11,680),
(4, 10,746), (5, 9886). Plot the points
in a coordinate plane, and draw a smooth curve
through the points.

   b. From the graph, the value of your truck in 6 years is about $9095.

Exercises for Example 2

4. Use the graph in Example 2 to estimate the value of your truck
   in 8 years.

5. Graph the exponential decay model in Exercise 3.

Algebra 1
Chapter 8 Resource Book

NAME _____ DATE _____

Quick Catch-Up for Absent Students

For use with pages 483–492

The items checked below were covered in class on (date missed) _____

Activity 8.6: Investigating Exponential Decay (p. 483)

____ **Goal:** Show that an exponential decay model is decreasing.

Lesson 8.6: Exponential Decay Functions

____ **Goal 1:** Write and use models for exponential decay. (pp. 484–485)

Material Covered:

____ Example 1: Writing an Exponential Decay Model

____ Student Help: Skills Review

____ Example 2: Writing an Exponential Decay Model

____ Example 3: Making a List to Verify a Model

Vocabulary:

exponential decay, p. 484   initial amount, p. 484

time period, p. 484   decay factor, p. 484

decay rate, p. 484   percent of decrease, p. 484

____ **Goal 2:** Graph models for exponential decay. (pp. 486–487)

Material Covered:

____ Example 4: Graphing an Exponential Decay Model

____ Example 5: Graphing an Exponential Decay Model

____ Example 6: Comparing Growth and Decay Models

Activity 8.6: Fitting Exponential Models (p. 492)

____ **Goal:** Find a best-fitting exponential growth or decay model using a graphing calculator.

____ Student Help: Keystroke Help

____ Other (specify) _____

_____

Homework and Additional Learning Support

____ Textbook (specify) _pp. 488–491_____

_____

____ *Reteaching with Practice* worksheet (specify exercises)_____

____ *Personal Student Tutor* for Lesson 8.6

NAME _____ DATE _____

Cooperative Learning Activity

For use with pages 484–491

GOAL **To make a study guide for the lessons you have learned in this chapter**

Materials: paper, pencil

Exploring Exponents

Exponents are used in many different fields of mathematical study. In this activity, you and your partner will create a study guide explaining the rules of exponents. This guide should be tailored for use by next year's algebra class.

Instructions

1 Search the chapter for topics to be covered in your study guide. You must include at least four different concepts.

2 Create the study guide. Each concept should be explained thoroughly. Include both the general rule and several examples. Practice exercises must also be part of the book, including original word problems. You may not copy ideas or exercises from the chapter.

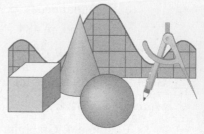

3 Create a cover for your study guide and bind the book.

Analyzing the Results

1. What was the hardest concept to learn in this chapter? Did you adequately explain this concept in your book?

2. Would an incoming algebra student be able to learn about exponents if he/she only had your book as a guide?

3. Is your book free of spelling and grammatical errors?

Real-Life Application: When Will I Ever Use This?

For use with pages 484–491

Record Albums

Listening to music is a popular leisure activity. The ever growing, billion dollar music recording industry has seen numerous changes over the past twenty-five years. Not only has the variety of music types widened to include such categories as Rap, Alternative, and Latin, but the rapid technological advances have also taken consumers through several recording media.

Prior to the 1980s, the standard recording medium was the vinyl LP. With the advent of the cassette, however, the LP suffered a quick decline. The smaller, more portable cassette tape became the norm. A car cassette player and a personal cassette player with headphones were considered standard equipment. With the introduction of the compact disc, or CD, the cassette tape did not remain on top for very long. Although, by 1992, the CD was the medium of choice, the decline of the cassette tape proved to be much more gradual than that of the vinyl LP. Without a doubt, the development of the high-density digital versatile disc, or audio DVD, will soon compete with the compact discs of today.

1. From 1987 to 1993, the decline of the vinyl LP can be defined by the exponential decay function $y = 134.3(0.45)^t$, where y is the millions of units sold and t is the number of years since 1987.

   a. At what rate did the sale of vinyl LP's decline each year?

   b. Estimate the number of vinyl LP's sold in 1993.

2. Graph the exponential decay model in Exercise 1.

3. The decline of the cassette tape and the growth of the compact disc were both linear models. Why do you think the transition from the cassette tape to the CD was more gradual than the shift from the vinyl LP to the cassette tape?

NAME _____ DATE _____

Math and History Application

For use with page 491

HISTORY   Seneca, a philosopher, used a globe of water to help him read and record very small letters around A.D. 65. Since then scientists have developed powerful and complex microscopes to make very small objects look larger so they can be seen and studied.

Zacharias Janssen invented the compound microscope about 1595. The microscope consisted of lenses attached at both ends of a central tube. A compound microscope contains at least two lenses, whereas a simple microscope has only one lens. One of the first books about a discovery using the microscope was *Micrographia,* written by Robert Hooke in 1665, in which he describes plant tissue viewed through a microscope.

Anthony van Leeuwenhoek was very successful at grinding lenses. About 1683 he made observations of blood with a single lens microscope with a magnification of 300 times. In the 1700s, the work of Chester Moor Hall and Benjamin Martin led to the development of the first achromatic microscope. This type of microscope eliminated color distortion that earlier microscopes had.

MATH   In the 1930s an electron microscope was developed. An electron microscope uses a beam of electrons instead of light. An electron microscope was built by Max Knoll and Ernst Ruska in Germany in the early 1930s. Vladimir Zworykin supervised the development of an electron microscope that was 10 feet high and weighed about 1000 pounds at the RCA Lab in New Jersey in 1939. An electron microscope magnifies about 1,000,000 times, whereas an optical microscope only magnifies 20,000 times. Electron microscopes allow scientists to see objects as small as viruses and protein chains.

1. Write the magnification of an electron microscope and an optical microscope in scientific notation.

2. How many times stronger is the magnification of the electron microscope compared to the optical microscope? Write your answer in scientific notation.

3. Write the measurements listed below in scientific notation.

    a. the diameter of a red blood corpuscle is 0.00001 gram

    b. a virus is 0.000000019 meter

    c. the diameter of a hydrogen molecule is 0.00000001 centimeter

NAME _____ DATE _____

Challenge: Skills and Applications

For use with pages 484–491

In Exercises 1–3, use the following information.

The pressure exerted by Earth's atmosphere at sea level is about 14.7 pounds per square inch. For every mile above sea level, the pressure decreases by about 18%

1. Write an equation for the pressure y (in pounds/square inch) at height x (in miles) above sea level.

2. Use a graph of the function from Exercise 1 to determine at what height the atmospheric pressure is half of what it is at sea level.

3. Find out how many more miles above the height from Exercise 2 the atmospheric pressure will be halved again.

In Exercises 4–5, use the following information.

If an element is radioactive, its chemical structure changes with time. The *half-life* of a radioactive element is the amount of time it takes for half of any amount of the element to undergo this change. Neptunium-232 is a radioactive form of an artificially produced element with a half-life of about 15 minutes.

4. Write an equation that relates the number of grams y of Neptunium-232 that remain after t minutes from a 10-gram sample.

   (Hint: $\dfrac{t}{15} = $ the number of 15-minute time periods in t minutes.)

5. About how long will it take before only about 5% of the original sample remains?

NAME _____ DATE _____

Quiz 2

For use after Lessons 8.4–8.6

1. Rewrite 8.67×10^{-7} in decimal form. *(Lesson 8.4)*

2. Rewrite 73,480,000 in scientific notation. *(Lesson 8.4)*

3. Evaluate the expression $(1.2 \times 10^{-4})(3 \times 10^{6})$. *(Lesson 8.4)*

4. You deposit $1000 in an account that pays 7.5% annual interest compounded yearly. What is the account balance after 5 years? Round your answer to the nearest cent. *(Lesson 8.5)*

5. A town has a population of 29,000. The population is decreasing by 2.5% each year. At this rate, what will the population be after 10 years? *(Lesson 8.6)*

Answers

1. _____

2. _____

3. _____

4. _____

5. _____

Chapter Review Games and Activities

For use after Chapter 8

Solve the following problems, and find the answer at the bottom of the page. Place the letter of the answer on the line with the problem number to answer the riddle.

What does the amateur magician who works as a mortician say when he pulls a rabbit out of his hat?

1. $x^3 \cdot x^4 \cdot x^2 =$

2. $3^{-3} \cdot 3^2 =$

3. $(x^2)^4 =$

4. $2x^6 \cdot (3x)^2 =$

5. $(3^{-2})^{-1} =$

6. $x^3 \cdot x^4 =$

7. $(-5)^0 \cdot x =$

8. $(-xy)(x^2y)^3 =$

9. $\dfrac{x^3 \cdot x^5}{x^6} =$

10. $\dfrac{2^{-4} \cdot 2^3}{2^{-2}} =$

11. $\dfrac{(x^3)^2}{(x^5)^3} =$

12. $(-3xy)^4(-x^3) =$

Answers:

(a) x^2 (c) x^7 (n) x^6 (e) x^{14} (m) x^{12} (b) x^8

(m) 3 (e) $\dfrac{1}{x^9}$ (v) 2 (l) x^{24} (a) x (r) $18x^8$

(a) 9 (t) $12x^4y^4$ (o) $\dfrac{1}{x^{10}}$ (a) x^9 (s) $-5x$ (m) $36x^8$

(f) $\frac{1}{6}$ (d) $-x^7y^4$ (r) $-81x^7y^4$ (b) $\frac{1}{3}$

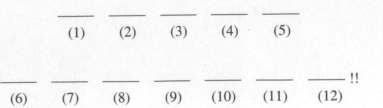

$\overline{}$ $\overline{}$ $\overline{}$ $\overline{}$ $\overline{}$
   (1)   (2)   (3)   (4)   (5)

$\overline{}$ $\overline{}$ $\overline{}$ $\overline{}$ $\overline{}$ $\overline{}$ $\overline{}$!!
(6)   (7)   (8)   (9)   (10)   (11)   (12)

Chapter Test A

For use after Chapter 8

Simplify the expression, if possible. Write your answer as a power.

1. $4^3 \cdot 4^5$

2. $(2^3)^4$

3. $(6 \cdot 7)^3$

4. $(4xy^2)^2$

Simplify. Then evaluate the expression when $a = 1$ and $b = 2$.

5. $b^3 \cdot b^4$

6. $(a^2)^3$

Complete the statement using < or >.

7. $(5 \cdot 3)^3$ ___?___ $5 \cdot 3^3$

8. $2^2 \cdot 4^6$ ___?___ $(2 \cdot 4)^6$

Evaluate the expression. Write your answer as a fraction in simplest form.

9. 5^{-3}

10. $\left(\dfrac{1}{3}\right)^{-1}$

Rewrite the expression with positive exponents.

11. y^{-3}

12. $\dfrac{1}{3x^{-3}}$

Match the equation with its graph.

A.

B.

C.

13. $y = 2^x$

14. $y = 3x$

15. $y = 3^x$

Evaluate the expression. Write your answer as a fraction in simplest form.

16. $\dfrac{2^5}{2^3}$

17. $\left(\dfrac{2}{3}\right)^3$

Answers

1. _____
2. _____
3. _____
4. _____
5. _____
6. _____
7. _____
8. _____
9. _____
10. _____
11. _____
12. _____
13. _____
14. _____
15. _____
16. _____
17. _____

Algebra 1
Chapter 8 Resource Book

Review and Assess

NAME _____ DATE _____

Chapter Test A

For use after Chapter 8

Simplify the expression. The simplified expression should have no negative exponents.

18. $\left(\dfrac{4}{x}\right)^3$

19. $\dfrac{y^8}{y^9}$

Rewrite the number in decimal form.

20. 6.15×10^2

21. 1.14×10^{-2}

Rewrite the number in scientific notation.

22. 0.02

23. 1042

Evaluate the expression without using a calculator. Write the result in decimal form.

24. $(3 \times 10^{-2}) \cdot (12 \times 10^3)$

25. $\dfrac{4 \times 10^{-2}}{2 \times 10^{-3}}$

26. In 1998, the population of a city was 100,000. Then each year for the next five years, the population increased by 3%. Write an exponential growth model to represent this situation.

27. You buy a used truck for $10,000. It depreciates at the rate of 18% per year. Find the value of the truck after 5 years.

18. _____

19. _____

20. _____

21. _____

22. _____

23. _____

24. _____

25. _____

26. _____

27. _____

28. _____

29. _____

30. _____

31. _____

32. _____

Match the equation with its graph.

A.

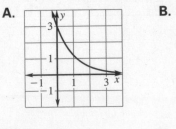

B.

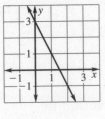

C.

28. $y = 3 - 2x$

29. $y = 3(1.4)^x$

30. $y = 3(0.4)^x$

Classify the model as exponential growth or exponential decay.

31. $y = 17(1.9)^x$

32. $y = 22(0.8)^x$

Review and Assess

Chapter Test B

For use after Chapter 8

Simplify the expression, if possible. Write your answer as a power.

1. $6^7 \cdot 6^9$

2. $(x^5)^6$

3. $(-2x)^4$

4. $3x^2 \cdot (4x^3)^2$

Simplify. Then evaluate the expression when $a = 1$ and $b = 2$.

5. $(a^2b^2)^3$

6. $a^2 \cdot (a^3b^2)^4$

Complete the statement using < or >.

7. $(8^3 \cdot 8^5)$ __?__ 8^{15}

8. $3^4 \cdot 3^5$ __?__ 3^{20}

Evaluate the expression. Write your answer as a fraction in simplest form.

9. $2(2^{-5})$

10. $\left(\frac{1}{4}\right)^{-2}$

Rewrite the expression with positive exponents.

11. $x^{-4}y^3$

12. $\dfrac{1}{2x^{-2}y^{-3}}$

Match the equation with its graph.

A.

B.

C.

13. $y = 4^x$

14. $y = \frac{1}{2}x$

15. $y = 5^x$

Evaluate the expression. Write your answer as a fraction in simplest form.

16. $\dfrac{4^5 \cdot 4^3}{4^2}$

17. $\left(\dfrac{7}{8}\right)^2$

1.	_____
2.	_____
3.	_____
4.	_____
5.	_____
6.	_____
7.	_____
8.	_____
9.	_____
10.	_____
11.	_____
12.	_____
13.	_____
14.	_____
15.	_____
16.	_____
17.	_____

NAME _____ DATE _____

Chapter Test B

For use after Chapter 8

Simplify the expression. The simplified expression should have no negative exponents.

18. $\left(\dfrac{4x^2y}{3xy^2}\right)^4$　　　　　**19.** $\left(\dfrac{2x^3y^2}{3xy}\right)^{-3}$

Rewrite the number in decimal form.

20. 4.3269×10^3　　　　　**21.** 7.1532×10^{-5}

Rewrite the number in scientific notation.

22. 0.0032　　　　　**23.** $321{,}562.5$

Evaluate the expression without using a calculator. Write the result in decimal form.

24. $(6 \times 10^{-2}) \cdot (7 \times 10^{-3})$　　　　　**25.** $\dfrac{5 \times 10^{-3}}{2 \times 10^{-4}}$

26. You deposit $2000 in an account that pays 8% interest compounded yearly. Find the balance of the account after 3 years.

27. A city had a declining population from 1992 to 1998. The population in 1992 was 200,000. Each year for 6 years, the population declined by 3%. Write an exponential decay model to represent this situation.

18. _____

19. _____

20. _____

21. _____

22 _____

23. _____

24. _____

25. _____

26. _____

27. _____

28. _____

29. _____

30. _____

31. _____

32. _____

Match the equation with its graph.

A. 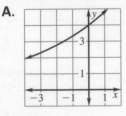　　**B.** 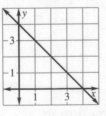　　**C.**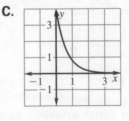

28. $y = 4 - x$　　　　**29.** $y = 4(1.2)^x$　　　　**30.** $y = 4(0.2)^x$

Classify the model as exponential growth or exponential decay.

31. $y = 11(4.25)^x$　　　　　　**32.** $y = 46(0.78)^x$

Review and Assess

Chapter Test C

For use after Chapter 8

Simplify the expression, if possible. Write your answer as a power.

1. $x^8 \cdot x^{10} \cdot x^4$

2. $[(x-1)^3]^8$

3. $(-3xy)^5$

4. $(-3a^2b^3)^4(2ab)^3$

Simplify. Then evaluate the expression when $a = 1$ and $b = 2$.

5. $-(a^5b^2)^3$

6. $(a^3b^4) \cdot (a^2b^3)^4$

Complete the statement using < or >.

7. $(9^3)^4$ __?__ 9^{11}

8. $(4^3 \cdot 7)^5$ __?__ $4^{14} \cdot 7^5$

Evaluate the expression. Write your answer as a fraction in simplest form.

9. $4^3 \cdot 0^{-2}$

10. $5^8 \cdot 5^{-8}$

Rewrite the expression with positive exponents.

11. $(8a^{-5})^2$

12. $\dfrac{1}{(4b^{-6})^2}$

Match the equation with its graph.

A.

B.

C.

13. $y = 2^x$

14. $y = \frac{2}{3}x$

15. $y = 4^x$

Evaluate the expression. Write your answer as a fraction in simplest form.

16. $\dfrac{(-5)^3}{-5^3}$

17. $\left(-\dfrac{4}{5}\right)^3$

1. _____

2. _____

3. _____

4. _____

5. _____

6. _____

7. _____

8. _____

9. _____

10. _____

11. _____

12. _____

13. _____

14. _____

15. _____

16. _____

17. _____

Simplify the expression. The simplified expression should have no negative exponents.

18. $\dfrac{10x^3y^2}{4xy^2} \cdot \dfrac{8x^5y^3}{3x}$

19. $\dfrac{8x^{-2}y^5}{9x^2y} \cdot \dfrac{3x^5y^{-3}}{4x^{-2}y}$

Rewrite the number in decimal form.

20. 8.92635×10^8

21. 1.6935×10^{-8}

Rewrite the number in scientific notation.

22. 0.0000159

23. $168,269.83$

Evaluate the expression without using a calculator. Write the result in decimal form.

24. $(8 \times 10^5) \cdot (1.2 \times 10^{-4})$

25. $\dfrac{8.8 \times 10^{-1}}{1.1 \times 10^{-1}}$

26. In 1998, the population of a city was 250,000. Then each year for the next five years, the population increased by 4.5%. Write an exponential growth model to represent this situation.

27. You buy a used truck for $14,000. It depreciates at the rate of 17% per year. Find the value of the truck after 3 years.

18. _____	
19. _____	
20. _____	
21. _____	
22. _____	
23. _____	
24. _____	
25. _____	
26. _____	
27. _____	
28. _____	
29. _____	
30. _____	
31. _____	
32. _____	

Match the equation with its graph.

A. 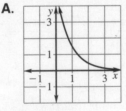 **B.** 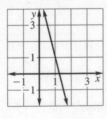 **C.**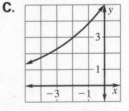

28. $y = 5 - 4x$

29. $y = 5(1.3)^x$

30. $y = 5(0.3)^x$

Classify the model as exponential growth or exponential decay.

31. $y = 12\left(\tfrac{8}{7}\right)^x$

32. $y = 18\left(\tfrac{4}{5}\right)^x$

Review and Assess

1. Simplify $(3^4 \cdot 3^3)^2$.

   (A) 3^9 (B) 3^{14}

   (C) 3^{16} (D) 3^{24}

2. Simplify $[(1 + x^2)]^3$ when $x = 2$.

   (A) 33   (B) 65

   (C) 125   (D) 721

3. What is the equation of the graph?

   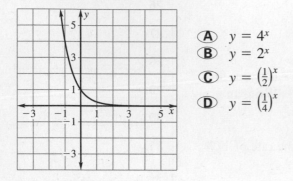

   (A) $y = 4^x$
   (B) $y = 2^x$
   (C) $y = \left(\frac{1}{2}\right)^x$
   (D) $y = \left(\frac{1}{4}\right)^x$

4. Simplify $\dfrac{3x^2y}{4x^3y^2} \cdot \dfrac{8x^{-3}y^5}{3x^{-5}y}$.

   (A) $\dfrac{2}{x^3y}$ (B) $\dfrac{x^3y}{2}$

   (C) $2x^3y$ (D) $2xy^3$

5. Which of the following numbers is *not* written in scientific notation?

   (A) 61.2×10^4 (B) 4.56×10^{18}

   (C) 8.642×10^{-3} (D) 1.1987×10^3

6. Rewrite 4.5×10^{-7} in decimal form.

   (A) 0.00000045   (B) 0.0000045

   (C) 45,000,000   (D) 450,000,000

7. Which of the models below are exponential decay models?

   I. $y = 1.19^x$ II. $y = 0.12^x$

   III. $y = \left(\frac{5}{3}\right)^x$ IV. $y = \left(\frac{2}{3}\right)^x$

   (A) I and II   (B) II and III

   (C) I and III   (D) II and IV

In Questions 8 and 9, choose the statement below that is true about the given numbers.

   A.   The number in column A is greater.

   B.   The number in column B is greater.

   C.   The two numbers are equal.

   D.   The relationship cannot be determined from the given information.

8.

	Column A	Column B
When $x = 0$,	$y = \left(\frac{2}{3}\right)^x$	$y = \left(\frac{2}{3}\right)^{-x}$

   (A)   (B)   (C)   (D)

9.

	Column A	Column B
When $x = -1$,	$y = 5x$	$y = 5^x$

   (A)   (B)   (C)   (D)

10. You deposit $500 in an account that pays 5% interest compounded yearly. How much money is in the account after 4 years?

   (A) $541.90   (B) $607.75

   (C) $1118.03   (D) $2531.25

11. You buy a used truck for $22,000. It depreciates at the rate of 10% per year. What is the value of the truck after 4 years?

   (A) $13,000   (B) $14,434.20

   (C) $21,133.11   (D) $21,912.13

JOURNAL

1. The following problems represent a quiz paper from a student in algebra. Some of the problems are incorrect. In this journal, you are to be the "teacher" and help the student to understand which parts are correct, and which parts are incorrect. *Do not* correct the student's work. Instead, write a sentence for each problem to tell the student why the result is correct (or incorrect) and explain why (or why not).

a. $(-3x^2)^3 = -3x^6$

b. $3(ab^3)^4 = 3a^4b^{12}$

c. $2a^2b^2 \cdot 4a^5b^2 = 8a^{10}b^4$

d. $\left(\dfrac{x}{y}\right)^{-2} = \dfrac{1}{x^2y}$

MULTI-STEP PROBLEM

2. The following data set represents the enrollment at Adlai E. Stevenson High School in Lincolnshire, Illinois, since the fall of 1980. Let x be the number of years since 1980.

Year, x	1980	1981	1982	1983	1984	1985	1986	1987	1988	1989
Enrollment, y	1414	1454	1439	1555	1617	1727	1717	1804	1833	1881

Year, x	1990	1991	1992	1993	1994	1995	1996	1997	1998
Enrollment, y	1965	2176	2449	2679	2885	2999	3124	3301	3503

a. Compare the enrollment for each year to the previous year and find the percent increase (or decrease) in enrollment. Round to the nearest hundredth.

b. Is the enrollment per year increasing or decreasing overall? By what percent?

c. Write an exponential model that approximates the data, where x is the number of years since 1980 and y is the school enrollment. Does the model represent exponential growth or exponential decay? Explain.

d. Predict the enrollment for the year 2008.

e. Repeat part (c) using your graphing calculator to find an exponential model. Compare this equation to the equation from part (c).

f. Use the equation from part (e) to estimate the year enrollment will be about 7875 students. Explain how this was determined.

3. *Writing* Why is it important for the school district to have a model that approximates enrollment? How would the model be useful?

Review and Assess

JOURNAL SOLUTION

1. a. The result is incorrect. When finding a power of a product, find the power of *each* factor and multiply.

    b. The result is correct. To find a power of a product, find the power of *each* factor and multiply.

    c. The result is incorrect. When multiplying powers with the same base (such as *a*), *add* the exponents.

    d. The result is incorrect. When finding the power of a quotient, find the power of the numerator *and* the power of the denominator.

MULTI-STEP PROBLEM SOLUTION

2. a. percents: 3%, −1%, 8%, 4%, 7%, −1%, 5%, 2%, 3%, 4%, 11%, 13%, 9%, 8%, 4%, 4%, 6%, 6%

    b. Enrollment is increasing overall by about 5% per year.

    c. $y = 1414(1.05)^x$; exponential growth; the model represents exponential growth because the enrollment increases about 5% each year.

    d. 5543 students

    e. $y = 1296(1.055)^x$; This model has a similar percent to the model found in part (c), but the *y*-intercept is different.

    f. The enrollment will be about 7875 in 2014. Students can use their graphing calculator's table feature to find this solution.

3. *Writing* Answers may vary. *Sample answer:* The school district needs this information in order to plan for the years to come. This would include supplies, teachers and staff members, as well as building space.

MULTI-STEP PROBLEM RUBRIC

4 Students complete all parts of the questions accurately. Explanations are logical and clear. Students correctly find an equation of best-fit by hand and by using the regression features of a calculator. The model used is exponential growth. Students are able to use their graphing calculator features to answer the questions.

3 Students complete the questions and explanations. Solutions may contain minor mathematical errors or misunderstandings. Students are able to use a graphing calculator to answer the questions.

2 Students complete questions and explanations. Several mathematical errors may occur. Explanations do not fit the questions. Students are unable to correctly use a graphing calculator to answer questions (e) and (f).

1 Students' work is very incomplete. Solutions and reasoning are incorrect. Students are unable to use a graphing calculator to answer (e) and (f).

Review and Assess

108 Algebra 1
Chapter 8 Resource Book

CHAPTER
8

Project: City Growth

For use with Chapter 8

OBJECTIVE Compare the growth of the city where you live to cities in other parts of the country.

MATERIALS paper, pencil, graphing calculator or computer (optional)

INVESTIGATION Different areas of the country grow at different rates. Which region do you think is growing the fastest? Do you think big cities grow faster than small towns?

Collect data on the population of five cities in 1970, 1980, 1990, and 2000. Include the closest city to where you live as well as one city in each of the following regions of the United States: south, northeast, midwest, and west coast. Include at least one large city (more than 2 million people), one small city (about a half million people), and a small town (less than 100,000 people).

1. Assuming exponential growth, use the 1970, 1980, and 1990 data to write an exponential equation for the population y of each city x decades after 1970.

2. Use your equations to predict the population of each city in the year 2000. Compare each to the actual data. For which cities does the equation give a good prediction?

3. Find a linear equation to fit the 1970, 1980, and 1990 data for the cities that did not exhibit exponential growth. Use the equations to predict the population in 2000. Compare your prediction to the actual data. Did the linear model give a better prediction for these cities than the exponential model did?

4. Draw a graph for each city using the equation that gave the best prediction. Plot the data points on the graph.

PRESENT YOUR RESULTS Write a report presenting your data. Analyze what your data implies for population growth in different regions of the country and for different size cities. How does the growth of your city fit the trends you found? Include the graphs of the best model for each city you investigated.

Review and Assess

Project: Teacher's Notes

For use with Chapter 8

GOALS • Write, use, and graph models for exponential growth.

• Determine whether a linear model is appropriate and use a linear model to make a real-life prediction.

• Use mathematical models to make predictions and analyze trends.

MANAGING THE PROJECT Encourage students to chose the cities carefully. You may want to require that students choose different cities in each region to provide a better basis for the discussion of trends at the end of the project. Students should be able to find data from the United States Census Bureau on the Internet, in almanacs, or from each city government, chamber of commerce, or web site.

RUBRIC The following rubric can be used to assess student work.

4 The choice of cities shows attention to variety. The student chooses the best model to fit each set of data, makes appropriate predictions, analyzes the accuracy of the predictions correctly, and draws the graphs to effectively show the trends. The report demonstrates insight into the trends of the data and makes a convincing argument to support generalizations made.

3 The student chooses five cities of different size and in different regions, collects the data, writes a linear or exponential model for each city's population, and draws the graphs. However, the student may not find all models correctly or may not choose the best type of model for each city. The report discusses trends but the supporting arguments may not be as convincing as possible.

2 The student chooses cities, collects data, writes models, and draws graphs. However, work may be incomplete or reflect misunderstanding. For example, the student may use only exponential models, even for linear trends, or some models may not fit the data. The report may indicate a limited grasp of key ideas such as what the data indicate about regional population growth or it may lack key supporting evidence.

1 Data, models, predictions, and analysis of city and regional trends may be missing or do not show an understanding of key ideas. The report does not give an analysis of overall trends or does not support the analysis given.

Cumulative Review

For use after Chapters 1–8

Evaluate the expression when $b = -4$. (1.3)

1. $7 - b + b^2$

2. $b(6 - b^3)$

3. $\dfrac{b - 4}{-b}$

4. $-4.5(b - 2b)$

5. $-\dfrac{b}{8}$

6. $b(5 - b) \div 10b$

Find the difference. (2.3)

7. $12(2.1) - 10$

8. $-6(9 - 6)$

9. $-\frac{1}{25}(90 - 5)(-1)$

10. $-8(6.3 - 1)$

11. $-10(4 - 4) + 5^2$

12. $(-6 + 2)3 - 3^2$

Solve the equation. (3.4)

13. $5x + 3(x + 4) = 28$

14. $66 = -\frac{6}{5}(x + 3)$

15. $-28 = 2(x + 3) - 5(x - 1)$

16. $6 = \frac{3}{2}x + \frac{1}{2}(x - 4)$

Find the slope of the line passing through the given points. (4.4)

17. $(2, 3), (4, -3)$

18. $(7, 0), (2, -2)$

19. $(-9, -5), (9, 8)$

20. $\left(-\frac{5}{6}, \frac{8}{3}\right), (1, 0)$

21. $(0, -1), (-1, 0)$

22. $(7, -2), (-7, 2)$

Find the constant of variation and the slope of the direct variation model. (4.5)

23. $y = 7x$

24. $y = \frac{6}{5}x$

25. $y = 0.2x$

26. $y = -x$

27. $y = x$

28. $y - 0.23x = 0$

Find the x-intercept and the y-intercept of the graph of the equation. (4.3)

29. $6x - 2y = -7$

30. $\frac{1}{4}x = 1 - y$

31. $-4x + 5y = -8$

32. $-y = -12 + 2x$

33. $0.2x - 7 = 12y$

34. $-\frac{8}{9}x - \frac{19}{3}y = -\frac{2}{3}$

Write an equation of the line in slope-intercept form. (5.1)

35. The slope is 7; the y-intercept is -6.

36. The slope is 1; the y-intercept is $\frac{6}{7}$.

37. The slope is $\frac{3}{4}$; the y-intercept is -2.

Find the mean, median, and mode of the collection of numbers. (6.6)

38. 2, 3, 2, 5, 1, 1, 5, 4, 10, 2

39. 6, 15, 9, 6, 52, 32, 8, 20, 26

NAME _____ DATE _____

Cumulative Review

For use after Chapters 1–8

Use substitution to solve the linear system. (7.2)

40. $2b = 16$
$a = b + 5$

41. $8x + 9y = 111$
$y = x + 1$

42. $x + \frac{3}{4}y = \frac{3}{4}$
$2y = 4x + 16$

43. $0.6x + 0.2y = 0.8$
$-x + y = 15$

Simplify, if possible. Write your answer as a power. (8.1)

44. $[(-3xy)^2]^4$

45. $[x^2(x-8)^3]^5$

46. $(-2a)^4 \cdot (-6a)^3$

47. $(-ab)(a^5b^9)^4$

48. $(5x^3y)^2 \cdot xy$

49. $(-10xy^4)^2\,(-4x^5y)$

Rewrite the expression with positive exponents. (8.2)

50. $(a^{-9}b^{-9})(a^3b^{-8})$

51. $-4^2 \cdot 6^{-1} \div 3^{-3}$

52. $\dfrac{1}{(8xy^{-6})^{-1}}$

53. $(-rs)^{-6}\left(\dfrac{45}{9r^8s^6}\right)$

54. $(-10a^3b^{-2})^0$

55. $\dfrac{rs^{11}}{r^{-9}}\,r^{-3}s^{-9}$

Simplify the expression. The simplified expression should have no negative exponents. (8.3)

56. $y\left(\dfrac{x^{-3}}{y^{-5}}\right)^{-1} \cdot x\left(\dfrac{y^3}{x^{-3}}\right)^{-3}$

57. $x^6 \cdot \dfrac{6}{(xy)^{-3}}$

58. $\dfrac{mn^{-9}m^{30}}{m^0m^{23}}$

59. $\left(\dfrac{18x^{-8}}{45x^{-9}}\right)$

60. $\dfrac{64r^3s^{-12}}{12r^{-9}s^{-12}} \cdot \dfrac{120r^6}{24s^{-18}}$

61. $\dfrac{88}{11xy} \cdot \dfrac{x}{(xy)^{-6}}$

Rewrite in scientific notation. (8.4)

62. 0.025

63. 93.5

64. 15

65. $99,002,356$

66. 632.02

67. 0.0004631

Find the balance after 10 years of an account that pays 5.7% interest compounded yearly given the following investment amounts. (8.5)

68. $389

69. $675

70. $6,004

Classify the model as *exponential growth* or *exponential decay*. Identify the growth or decay factor and the percent increase or decrease per time period. (8.6)

71. $y = 34(0.23)^t$

72. $y = 63(0.002)^t$

73. $21\left(\frac{12}{11}\right)^t$

74. $y = 237(0.56)^t$

75. $y = 5\left(\frac{8}{15}\right)^t$

76. $y = 8(1.11)^t$

Algebra 1
Chapter 8 Resource Book

ANSWERS

Chapter Support

Parent Guide
Chapter 8

8.1: second offer; the heir gets $1,048,575 with the second offer   **8.2:** notice the zero exponent first; 1   **8.3:** about 1.4 to 1   **8.4:** 450,000
8.5: about $712.61   **8.6:** about $8,995

Activity: $\frac{1}{4}$; $\left(\frac{1}{2}\right)^t$ or $\left(1 - \frac{1}{2}\right)^t$; no

Prerequisite Skills Review

1. $\left(\frac{1}{2}\right)^2$ **2.** -3^3 **3.** $(5x)^3$ **4.** $(82)^2$
5. 2   **6.** -256 **7.** $\frac{3}{4}$ **8.** $\frac{1}{9}$ **9.** -2
10. -2 **11.** -49 **12.** 3

13. 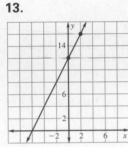   **14.**

15. 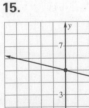   **16.**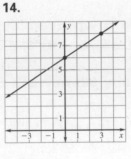

Strategies for Reading Mathematics

Answers may vary. Examples given.

1. exponent is x; base is b;

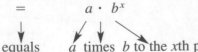

2. a times the sum of b and c equals the sum of the product a times b and the product a times c; Check diagrams or examples. Students may use an area model for multiplication and show that a rectangle with dimensions a and $(b + c)$ can be split into two smaller rectangles, one with dimensions a and b and one with dimensions a and c; The distributive property shows that when you multiply the terms of an addition expression by a factor, you can multiply each term separately by the factor and then add the two products together.

3. V equals s to the third power; Check diagrams or examples. A diagram can be used to show that a cube with side length s has a base with area $s \cdot s$ and a height s so its volume is $s \cdot s \cdot s = s^3$; You can find the volume V of a cube by taking the side length s as a factor three times.

Lesson 8.1

Warm-up Exercises
1. 64   **2.** 100   **3.** 256   **4.** 16

Daily Homework Quiz

1. 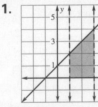   **2.**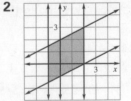

3. Check students' graphs;
$y \leq \frac{3}{4}x + \frac{7}{4}$; $y \geq 1$; $x \leq 3$

Lesson Opener
Allow 15 minutes.
The following boxes can be shaded: 8^5; $8^4 \cdot 8$; $(5 + 3)^5$; $8^2 \cdot 8^3$; $(4 \cdot 2)^5$; $8^5 \cdot 8^0$; $(4 + 4)^5$; $(2 \cdot 4)^5$; $2^5 \cdot 4^5$; $2^5 \cdot (2 \cdot 2)^5$; $2^5 \cdot 2^5 \cdot 2^5$; $(2 \cdot 2 \cdot 2)^5$; $(2^3)^5$; $2^7 \cdot 2^8$; 2^{15}; $2^{10} \cdot 2^5$

Practice A

1. 2^2 2. x^3 3. 3^3 4. n^5 5. m^7 6. 4^7

7. y^9 8. t^4 9. 2^5 10. 8 11. -27

12. -1 13. 256 14. 81 15. 256 16. x^{10}

17. y^{12} 18. $8x^3$ 19. 3^6 20. 2^{15} 21. x^8

22. y^{16} 23. $8x^3$ 24. $9x^8$ 25. x^{14} 26. $8x^5$

27. $x^3y^3z^{12}$ 28. x^6 29. $-4y^5$ 30. $5a^6$

31. x^2y, 4 32. x^5, 32 33. y^6, 1 34. x^4y, 16

35. x^4y^6, 16 36. $-x^8y^{12}$, -256 37. 256 ft^3

38. $V \approx 22.11x^2$ 39. 1,073,741,824

Practice B

1. 4^3 2. n^4 3. 2^5 4. x^6 5. 5^6 6. c^5

7. t^8 8. m^8 9. x^{10} 10. 16 11. -125

12. 36 13. $27g^3$ 14. a^2b^2 15. h^3t^6

16. x^{30} 17. y^{21} 18. $x^{18}y^9$ 19. 4^5 20. 9^{10}

21. $16a^2$ 22. 36^3, or $(-6)^6$ 23. $-6^{21}x^{42}y^{21}$

24. $(3x - 2)^9$ 25. $-4^3 \cdot 5^4 \cdot x^7$ 26. $256a^{11}b^2$

27. $\frac{1}{16}x^8$ 28. $a^8b^{10}c^{15}$ 29. x^{16} 30. $-2x^{11}y^7$

31. y^6, 64 32. x^4, 16 33. $-x^5$, -32

34. x^3y^9, 4096 35. $-x^6y^2$, -256

36. x^4y^5, 512

37. $4^{17} = 17,179,869,180$; $\dfrac{1}{17,179,869,180}$

38. $53,687,091.20 39. $(5x)^2$, 1600 mi^2

Practice C

1. x^5 2. 3^5 3. y^{10} 4. z^{17} 5. 6^{11} 6. t^9

7. $16x^2$ 8. $25x^4$ 9. $8t^6$ 10. m^4n^{10}

11. $16w^{12}$ 12. $-27y^6$ 13. 2^8 14. 8^6

15. $4x^4y^6$ 16. $-3^5a^2b^{12}c^{29}$ 17. $\frac{1}{8}x^3$ 18. $\frac{1}{9}x^8$

19. $\frac{81}{16}x^{18}$ 20. $-4^2 \cdot 3^3 \cdot y^8$ 21. $(9x + 15)^{18}$

22. $64x^{40}$ 23. $-a^{19}b^{29}$ 24. $r^9s^{31}t^{20}$

25. x^4y^7, 16 26. $-8x^3y^3$, -64 27. $\frac{3}{2}x^2y^3$, 6

28. $5x^2y^7$, 20 29. $144y^8$, 144

30. $-432x^3y^6$, -3456 31. $262,144 32. 4^{15}

33. 2^{10} 34. $4^{15} \cdot 2^{10}$, or 2^{40} 35. $\dfrac{1}{4^{15} \cdot 2^{10}}$, or $\dfrac{1}{2^{40}}$

Reteaching with Practice

1. m^2 2. 6^5 3. y^7 4. 3^6 5. w^{21} 6. 7^{15}

7. t^{12} 8. $(-2)^6$ 9. $125x^3$ 10. $100s^2$

11. x^4 12. $-27y^3$ 13. $\frac{9}{1}$

Real-Life Application

1. a. $8 \cdot 10^2$ b. $8 \cdot 10^2$ c. 10^4

2. a. $8 \cdot 10^2 \cdot 10^4 = 8 \cdot 10^6 = 8,000,000$

b. $8 \cdot 10^2 \cdot 8 \cdot 10^2 \cdot 10^4 = 8^2 \cdot 10^8 =$ 6,400,000,000

Challenge: Skills and Applications

1. k^{n+3} 2. x^{6n+2} 3. y^{21m} 4. k^{12mn}

5. $x^{6n}y^{2mn}$ 6. $x^{36mn+7}y^{54mn}$ 7. a^2b 8. 2

9. 25, 49 10. 133, 343 11. 256, 256

12. 2, 128 13. 8, 8 14. 54, 216

15. 100,000; 100,000 16. 11, 11

17. true only if (1) $a = 0$, $b = 0$, or both or (2) $n = 1$ 18. $2^8 \cdot 4^{12}$ 19. 2^{32}

Lesson 8.2

Warm-up Exercises

1. 16 2. -8 3. $\frac{1}{8}$ 4. 2.205 5. 2187

Daily Homework Quiz

1. 5^{11}, or 48,828,125 2. 3^{20}, or 3,486,784,401

3. 7^6m^6, or 117,649m^6 4. $(-4)^3x^{21}y^6$, or $-64x^{21}y^6$ 5. $(5p + 2)^{10}$ 6. $\left(\frac{2}{3}\right)^3 \cdot 5^4 \cdot h^{20}$, or $\frac{5000}{27}h^{20}$ 7. $(-6)^2 \cdot 9^3 \cdot s^{12}t^{14}$, or 26,244$s^{12}t^{14}$

8. a^4b^5; -2592 9. $<$

Lesson Opener

Allow 10 minutes.

1. $A(2, 4)$; $B(1, 2)$; $C(0, 1)$; $D\left(-1, \frac{1}{2}\right)$; $E\left(-2, \frac{1}{4}\right)$

2. It is greater than 1, and as the x-values get larger, the y-values become infinitely large.

3. The y-intercept is 1. 4. It is between 0 and 1, and as the x-values get smaller, the y-values get closer to 0.

Practice A

1. 4, 2, 1, $\frac{1}{2}$, $\frac{1}{4}$, $\frac{1}{8}$ 2. 9, 3, 1, $\frac{1}{3}$, $\frac{1}{9}$, $\frac{1}{27}$

3. 16, 4, 1, $\frac{1}{4}$, $\frac{1}{16}$, $\frac{1}{64}$ 4. $\frac{1}{27}$ 5. $\frac{1}{32}$ 6. 1 7. $\frac{1}{8}$

8. 3 9. 25 10. $\frac{1}{64}$ 11. $\frac{1}{36}$ 12. $-\frac{1}{8}$

13. $\dfrac{1}{x^8}$ 14. $\dfrac{3}{x^5}$ 15. $7x^2$ 16. $9x^4$ 17. $\dfrac{8}{x^7y^8}$

Lesson 8.2 *continued*

18. $\dfrac{3}{a^3}$ **19.** $3y^3$ **20.** $\dfrac{1}{16x^2}$ **21.** $\dfrac{1}{16x^4}$ **22.** $\dfrac{1}{y^2}$

23. $27x^3$ **24.** $\dfrac{3y^5}{4x^2}$ **25.** $\dfrac{1}{27}, \dfrac{1}{9}, \dfrac{1}{3}, 1, 3, 9, 27$

26.

27. *y*-values increase

28. $8, 4, 2, 1, \dfrac{1}{2}, \dfrac{1}{4}, \dfrac{1}{8}$

29.

30. *y*-values decrease

Practice B

1. $\dfrac{1}{125}$ **2.** 3 **3.** $\dfrac{1}{216}$ **4.** -16 **5.** $\dfrac{1}{9}$

6. undefined **7.** 1 **8.** $\dfrac{1}{16}$ **9.** $\dfrac{1}{64}$ **10.** $\dfrac{1}{81}$

11. -64 **12.** 1 **13.** $\dfrac{4}{x^2}$ **14.** $\dfrac{x^4}{3}$ **15.** $\dfrac{x^3}{y^6}$

16. $\dfrac{7}{x^5 y}$ **17.** $\dfrac{x^2 y^7}{11}$ **18.** $\dfrac{1}{y^2}$ **19.** $\dfrac{1}{6561x^4}$

20. $\dfrac{y^{24}}{8x^9}$ **21.** $\dfrac{1}{128x^{70}}$ **22.** $3y^3$ **23.** $512x^6$

24. $\dfrac{1}{81}$ **25.** $\dfrac{9}{4}, \dfrac{3}{2}, 1, \dfrac{2}{3}, \dfrac{4}{9}$

26.

27. *y*-values decrease

28. 1990: ≈ 1206 **29.** 1960: ≈ 1869
 1995: $= 1200$ 1970: ≈ 1933
 2000: ≈ 1194 1980: $= 2000$
 2010: ≈ 1182 1990: ≈ 2069

Practice C

1. $\dfrac{1}{144}$ **2.** $\dfrac{125}{8}$ **3.** $\dfrac{1}{64}$ **4.** 27 **5.** 216
6. undefined **7.** 1 **8.** $\dfrac{1}{81}$ **9.** $\dfrac{1}{125}$ **10.** $\dfrac{1}{1000}$

11. -243 **12.** 0 **13.** $\dfrac{14}{x^5}$ **14.** $100x^7$ **15.** $\dfrac{y^{21}}{x^{10}}$

16. $\dfrac{20}{x^8 y^8}$ **17.** $\dfrac{x^3}{3y^9}$ **18.** $\dfrac{1}{121}$ **19.** $\dfrac{2401}{x^{16}}$

20. $\dfrac{x^{20} y^{60}}{1024}$ **21.** $-\dfrac{12y^6}{13x^{15}}$ **22.** $\dfrac{x^4}{4}$ **23.** $\dfrac{144y^4}{x^4}$

24. $-\dfrac{1}{100{,}000}$ **25.** $6.25, 2.5, 1, 0.4, 0.16$

26.

27. *y*-values decrease

28. 1990: ≈ 1206 **29.** 1960: ≈ 1869
 1995: $= 1200$ 1970: ≈ 1933
 2000: ≈ 1194 1980: $= 2000$
 2010: ≈ 1182 1990: ≈ 2069

30. $100, 50, 25, 12.5, 6.25, 3.125, 1.5625$

31. 1990: $\approx \$305.01$
 2000: $= \$600.00$
 2010: $\approx \$1180.29$

Reteaching with Practice

1. 1 **2.** 2 **3.** $\dfrac{1}{13^x}$ **4.** $\dfrac{1}{13y}$ **5.** $16x^4$

6. $\dfrac{d}{16c^4}$ **7.** 1 **8.** 16 **9.** $\dfrac{1}{625}$

10.

Lesson 8.2 *continued*

Interdisciplinary Application

1.

t	0	1	2	3	4	5
$A(t)$	100	50	25	12.5	6.25	3.125

2.

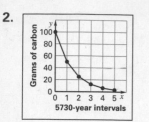

3. Answers may vary. *Sample answer:* The amount of carbon-14 left in an object after 50,000 or 60,000 years would be so small that detection would be impossible.

Challenge: Skills and Applications

1. negative **2.** positive **3.** positive

4. negative **5.** $\frac{16}{49}$ **6.** $\frac{1}{40}$ **7.** 1,000,000

8. $\frac{4}{25}$ **9.** \$1904.60 **10.** \$2809.47

11. \$1905.73 **12.** \$2457.92 **13.** \$2235.88

14. \$2226.76 **15.** \$2315.26 **16.** *Sample answer:* Compounding more frequently increases the value of the account slightly; the interest rate affects the value of the account much more than the frequency of compounding does. In this case, compounding six times as frequently only increased the value of the account about \$9 over the 8 years, whereas raising the interest rate by half a percent increased it by close to \$90.

Lesson 8.3

Warm-up Exercises

1. $\frac{1}{25}$ **2.** $9w^4$ **3.** $16v^2w^6$ **4.** $64x^6y^{17}$

5. $-32r^{10}s^{15}$

Daily Homework Quiz

1. $\frac{1}{49}$ **2.** 9 **3.** $\frac{1}{256}$ **4.** 1 **5.** $\frac{n^5}{m^3}$ **6.** $\frac{k^3}{7}$

7. 1 **8.** $1024x^3$ **9.**

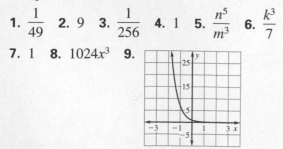

Lesson Opener

Allow 15 minutes.

1. first row: 8, 8; second row: 27, 27; third row: 64, 64 **2.** $\frac{a^m}{a^n} = a^{m-n}$ **3.** first row: 64, 64; second row: 9, 9; third row: 4, 4

4. $\left(\dfrac{a}{b}\right)^m = \dfrac{a^m}{b^m}$

Graphing Calculator Activity

1. $\frac{16}{25}$ **2.** $\frac{125}{27}$ **3.** $\frac{1}{16}$ **4.** $-\frac{125}{216}$ **5.** $\frac{4}{49}$ **6.** $\frac{81}{16}$

Practice A

1. 4 **2.** $\frac{1}{25}$ **3.** x^4 **4.** a^5 **5.** $\frac{1}{64}$ **6.** $\frac{1}{c^2}$

7. 4 **8.** -1 **9.** $\frac{1}{16}$ **10.** $\frac{8}{27}$ **11.** $\frac{16}{x^2}$

12. $\frac{9}{16}$ **13.** $\frac{27}{m^3}$ **14.** $\frac{x^4}{25}$ **15.** $\frac{64}{27}$ **16.** $\frac{a^{12}}{b^8}$

17. 49 **18.** $\frac{1}{36}$ **19.** 1 **20.** -1 **21.** 128

22. 16 **23.** $\frac{8}{27}$ **24.** $\frac{2}{3}$ **25.** $\frac{x^4}{81}$ **26.** x^5

27. $\frac{64}{x^6}$ **28.** $\frac{1}{x^3}$ **29.** x^9 **30.** $\frac{1}{x^2}$ **31.** y^4

32. $\frac{1}{m}$ **33.** 1 **34.** $\frac{16z^2}{3}$ **35.** $\frac{2a^4}{b^3}$ **36.** $\frac{27m^3n^3}{8}$

37. 2, 2.4, 2.88, 3.456 **38.** 1.44 to 1

Practice B

1. 16 **2.** $\frac{1}{64}$ **3.** x^6 **4.** $\frac{1}{b^4}$ **5.** y^6 **6.** 81

7. $\frac{1}{216}$ **8.** x **9.** $\frac{1}{81}$ **10.** $\frac{25}{36}$ **11.** $\frac{1024}{x^5}$

12. $\frac{y^3}{27}$ **13.** $\frac{25}{49}$ **14.** $\frac{64}{a^{15}}$ **15.** $\frac{x^{48}}{y^{24}}$ **16.** $\frac{c^{28}}{d^{40}}$

17. 32 **18.** -1 **19.** 125 **20.** 1 **21.** $\frac{1}{27}$

22. $\frac{49}{36}$ **23.** 64 **24.** $\frac{9}{64}$ **25.** $\frac{32}{x^5}$ **26.** x^{12}

27. $\frac{1}{b^{14}}$ **28.** $\frac{1}{r^3}$ **29.** $\frac{1}{t^{24}}$ **30.** a^{20} **31.** $\frac{343y^{18}}{x^{30}}$

32. $-\frac{25y}{x^2}$ **33.** $\frac{625}{81x^8y^{28}}$ **34.** $\frac{9y^4}{4x^7}$ **35.** $4x^3y$

Lesson 8.3 *continued*

36. $\dfrac{27y^8}{8x^9}$ **37.** $100, \approx 51.3, \approx 26.3, \approx 13.5,$
$\approx 6.9, \approx 3.5, \approx 1.8$ **38.** ≈ 7.4 to 1

Practice C

1. $\dfrac{1}{x^7}$ **2.** -125 **3.** $\dfrac{1}{243}$ **4.** x^2 **5.** $\dfrac{49}{9}$

6. $\dfrac{x^5}{16}$ **7.** $\dfrac{x^{36}}{y^{63}}$ **8.** $\dfrac{a^{100}}{b^{56}}$ **9.** 125 **10.** 49

11. $10{,}000$ **12.** -729 **13.** $\frac{1}{1024}$ **14.** $\frac{25}{729}$

15. 81 **16.** $-\dfrac{1}{32}$ **17.** $\dfrac{2x^7}{5}$ **18.** y^{56} **19.** $\dfrac{1}{a^6}$

20. $\dfrac{1331x^6}{y^9}$ **21.** $\dfrac{729}{64x^{54}y^{24}}$ **22.** x^{30} **23.** $-9x^2y^2$

24. $-\dfrac{2x^2}{5y}$ **25.** $\dfrac{3y^6}{x}$ **26.** $10x^{10}y^4$ **27.** $\dfrac{10y^3}{3x^{13}}$

28. $\dfrac{65x^2}{32y^4}$ **29.** $-\dfrac{4y^3}{x^3}$ **30.** $\dfrac{y^4}{x^2}$ **31.** $\dfrac{32y^6}{3x^6}$

32. ≈ 1.03 to 1 **33.** ≈ 1.48 to 1 **34.** $\frac{3125}{7776}$

Reteaching with Practice

1. $10^3 = 1000$ **2.** $\dfrac{1}{3}$ **3.** y^6 **4.** $\dfrac{16}{x^{12}}$ **5.** z^4

6. $\dfrac{25y^4}{w^2}$ **7.** $\left(\dfrac{1}{2}\right)^6 = \dfrac{1}{2^6} = \dfrac{1}{64} \approx 0.0156$

Real-Life Application

1. a. 7.7 hours per person per year

b. 128.4 hours per person per year

2. 16.7 to 1 **3.** No; because there are a limited number of hours that people are free to be on-line, the number of hours cannot increase exponentially for very long without eventually leveling off.

Challenge: Skills and Applications

1. 3 **2.** 3 **3.** -3 **4.** Let $n = -k$, where k is a positive integer. $(a^m)^n = (a^m)^{-k} =$
$\dfrac{1}{(a^m)^k} = \dfrac{1}{(a^{mk})} = a^{-mk} = a^{m(-k)} = a^{mn}$
5. Let $m = -j$ and $n = -k$, where j and k are positive integers. $(a^m)^n = (a^{-j})^{-k} = \dfrac{1}{(a^{-j})^k} =$

$\dfrac{1}{\left(\dfrac{1}{a^j}\right)^k} = \dfrac{1}{\left(\dfrac{1}{a^{jk}}\right)} = a^{jk} = a^{(-m)(-n)} = a^{mn}$

6. $(a^m)^n = (a^m)^0 = 1 = a^0 = a^{m \cdot 0} = a^{mn}$

7. The remainder is 1 each time.

8. The remainder is 1 each time. **9.** If p is a prime number and a is not divisible by p, then a^{p-1} leaves a remainder of 1 when divided by p.

Quiz 1

1. $72x^8y^7$ **2.** 3^4 **3.** 10 **4.** 216 **5.** $-\dfrac{1}{125y^3}$

6. $4y$ **7.** $\dfrac{-8}{125}$ **8.** $-\dfrac{6y^2}{x}$

Lesson 8.4

Warm-up Exercises

1. 1000 **2.** 0.0001 **3.** 10,000

4. 0.00000001 **5.** 10,000,000,000

Daily Homework Quiz

1. 49 **2.** $\dfrac{81}{625}$ **3.** $\dfrac{4}{9}$ **4.** b^8 **5.** $-\dfrac{m^3}{3n}$

6. $\dfrac{-135y^{10}}{x^2}$ **7.** 1.047^2

Lesson Opener

Allow 10 minutes.

Yes.

$3.6 \times 10^7 = 3.6 \times 10{,}000{,}000$
$\qquad\qquad = 36{,}000{,}000$

$6.72 \times 10^7 = 6.72 \times 10{,}000{,}000$
$\qquad\qquad = 67{,}200{,}000$

$9.275 \times 10^7 = 9.275 \times 10{,}000{,}000$
$\qquad\qquad = 92{,}750{,}000$

$1.413 \times 10^8 = 1.413 \times 100{,}000{,}000$
$\qquad\qquad = 141{,}300{,}000$

$4.836 \times 10^8 = 4.836 \times 100{,}000{,}000$
$\qquad\qquad = 483{,}600{,}000$

$8.87 \times 10^8 = 8.87 \times 100{,}000{,}000$
$\qquad\qquad = 887{,}000{,}000$

$1.78 \times 10^9 = 1.78 \times 1,000,000,000$
$= 1,780,000,000$
$2.794 \times 10^9 = 2.794 \times 1,000,000,000$
$= 2,794,000,000$
$3.658 \times 10^9 = 3.658 \times 1,000,000,000$
$= 3,658,000,000$

Practice A

1. 1000 **2.** 100,000 **3.** 0.01 **4.** 100

5. 0.001 **6.** 0.1 **7.** 10,000 **8.** 0.00001

9. 2030 **10.** 34,578 **11.** 64.3 **12.** 720,000

13. 5.2 **14.** 0.0468 **15.** 0.0000013

16. 0.008497 **17.** 0.00098 **18.** 2.5×10^4

19. 3.641×10^1 **20.** 4.0×10^6 **21.** 5.642×10^5

22. 9.32×10^0 **23.** 1.5×10^{-1} **24.** 8.3×10^{-3}

25. 7.18×10^{-7} **26.** 6.73×10^{-2}

27. 600,000,000,000 **28.** 0.000000009

29. 600 **30.** 0.8 **31.** 12,000,000,000

32. 0.0035 **33.** 24,000 **34.** 0.036 **35.** 0.42

36. 2.7×10^9 **37.** 1.86×10^5 **38.** 7.392×10^6

39. 2.86×10^{-3} **40.** 9.54×10^7 **41.** 3×10^{-2}

Practice B

1. 379,000 **2.** 0.025 **3.** 84.4 **4.** 65,393

5. 0.003589 **6.** 9.1187 **7.** 0.000010056

8. 726,587,460 **9.** 433,652,157,000

10. 6.43×10^1 **11.** 6.02×10^{-2}

12. 9.98653×10^2 **13.** 2.20786×10^7

14. 4.5668×10^1 **15.** 6.3×10^7

16. 7.485×10^{-3} **17.** 5.6388×10^{-6}

18. 7.96×10^9 **19.** 15,000 **20.** 0.00036

21. 2728 **22.** 6250 **23.** 6,000,000,000

24. 0.007 **25.** 2500 **26.** 0.000125

27. 4,410,000 **28.** 2.79646×10^9

29. 3.125×10^{-4} **30.** 8.5×10^8

31. 6.5928×10^6 **32.** 7.7×10^5

33. 7.56×10^{-3} **34.** 554,800,000; 5.548×10^8

35. $\approx 1.91 \times 10^2$

Practice C

1. 9,990,000 **2.** 0.001356 **3.** 60.1

4. 70,056.3 **5.** 0.00032 **6.** 3.50086

7. 77,889,523,314.6 **8.** 0.000000206230

9. 86,235,140,000,000 **10.** 3.2×10^{-2}

11. 3.3254236×10^4 **12.** 8.7×10^4

13. 2.22×10^{-5} **14.** 9.9578×10^{-4}

15. 7.0×10^7 **16.** 8.52246×10^{-1}

17. 8.0×10^{-9} **18.** 1.284×10^{10}

19. 0.000000072 **20.** 480,000 **21.** 3,400,000

22. 0.84 **23.** 12,400,000,000

24. 0.00000378 **25.** 3,750,000 **26.** 420,000

27. 0.0008 **28.** 16,000,000 **29.** 0.000000008

30. 4096 **31.** 1×10^{12} **32.** 7.25×10^{-4}

33. 1.2719×10^9 **34.** 6.681822×10^{-24} g

35. $\approx 2.91 \times 10^{-1}$

Reteaching with Practice

1. 9,332,000 **2.** 0.278 **3.** 450,000

4. 7.52×10^1 **5.** 1.35667×10^5

6. 8.8×10^{-4} **7.** 1.265×10^3

8. 8.0×10^{-3} **9.** $\approx 4.0 \times 10^2$

Cooperative Learning Activity

1. Pluto is $5.842 \cdot 10^9$ km farther from the sun than Mercury. **2.** Pluto is approximately $1.02 \cdot 10^2$ times farther from the sun than Mercury. **3.** Jupiter is approximately $5.2 \cdot 10^0$ times farther from the sun than the Earth.

Interdisciplinary Application

1. 1.6896×10^7; 5.808×10^6 **2.** 9.81×10^{13} square feet **3.** 2.45×10^{13} square feet

4. 4.9×10^{15} cubic feet

5. about 3.77×10^{24} grains of sand

Challenge: Skills and Applications

1. 356,600,000 **2.** 0.000000494

3. 390,000,000 **4.** 3.95×10^9

5. $3.95 \times 10^{n+5}$ **6.** 1.25×10^2, 5,000,000 times **7.** about 1.3 million times

8. about 1400 times

Lesson 8.5

Warm-up Exercises
1. 17.5 2. 288.29151 3. 33 4. 59,037

Daily Homework Quiz
1. 0.0000632 2. 4.55 3. 71,680,000
4. 3.59176×10^3 5. 8.49×10^{-5}
6. 4.6×10^8 7. 0.000000054 8. 75,000

Lesson Opener
Allow 10 minutes.
1. D; The growth can be given by
$700(1.02)(1.02)(1.02)(1.02)(1.02) = 700(1.02)^5$
2. B; The growth can be given by
$50,000(1.05)(1.05)(1.05) = 50,000(1.05)^3$
3. A; The growth can be given by
$100,000(1.03)(1.03)(1.03)(1.03) = 100,000(1.03)^4$.
4. *Sample answer:* In each answer, one plus the percent growth is raised to a power (the time period) and then multiplied by the beginning amount.

Graphing Calculator Activity
1. double 2. 18 years 3. 9 years
4. 6 years 5. 7.2 years

Practice A
1. 1.03 2. 1.09 3. 1.068 4. 1.102
5. 1.0101 6. 1.008 7. ≈$1462.32
8. ≈$1644.91 9. ≈$1924.32
10. ≈$3465.59 11. ≈$644.24 12. ≈$837.51
13. ≈$1288.48 14. ≈$1610.60 15. a 16. b
17. $T = 4000(1.08)^t$ 18. $P = 11,000(1.15)^t$
19.

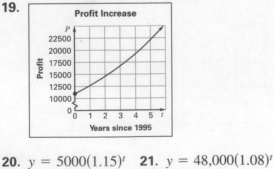

20. $y = 5000(1.15)^t$ 21. $y = 48,000(1.08)^t$
22. $y = 15(1.2)^t$

Practice B
1. ≈$1736.44 2. ≈$1914.42
3. ≈$2216.18 4. ≈$2443.34
5. ≈$644.24 6. ≈$708.67 7. ≈$773.09
8. ≈$1030.79 9. a 10. a
11. $P = 100,000(1.015)^t$
12. $P = 5,700,000(1.32)^t; 39,801,000$
13. $y = 5000(1.15)^t; \approx \$10,057$
14. $y = 48,000(1.08)^t; \approx 60,466$
15. $y = 15(1.2)^t; \approx 45$ words
16. $y = 50(1.3)^t; \approx 143$

Practice C
1. ≈$1513.52 2. ≈$1832.59 3. ≈$2218.93
4. ≈$2686.71 5. ≈$1366.73 6. ≈$2353.81
7. ≈$4555.76 8. ≈$6833.65 9. $10,718.59
10. $P = 375,000(1.0225)^t$
11. a.

t	1	2	3	4	5	6
A	2160	2332.8	2519.4	2721	2938.7	3173.7

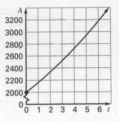

b.

x	1	2	3	4	5	6
y	3692.5	3895.6	4109.8	4335.9	4574.4	4825.9

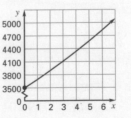

12. $y = 5000(1.125)^t; \approx \$9010$
13. $y = 48,000(1.083)^t; \approx 60,971$
14. 100% ; 800
15. 200% ; 3645

Reteaching with Practice

1. $472.78 **2.** $927.42 **3.** 1350 pheasants

4.
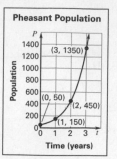

Pheasant Population

(3, 1350)
(0, 50)
(2, 450)
(1, 150)

Real-Life Application

1. about $2805.68 **2.** about $2981.54

3. about $4187.97

4.
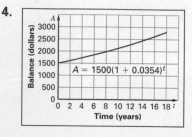

$A = 1500(1 + 0.0354)^t$

5.
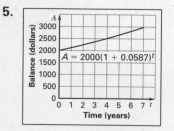

$A = 2000(1 + 0.0587)^t$

t	1	2	3	4
A	2000	2117	2242	2513

t	5	6	7
A	2660	2816	2982

Challenge: Skills and Applications

1. $y = 875,538(1.269)^t$ **2.** about 1,789,204

3. about 428,440 **4.** $y = C(1 + r)^t$

5. about 20,000 **6.** about 60,000

7. about 14.3 **8.** about 4.2 **9.** 523.4 Hz

10. 659.6 Hz **11.** 880.8 Hz

Lesson 8.6

Warm-up Exercises

1. 0.44 **2.** 0.89 **3.** 0.13 **4.** 0.89 **5.** 0.61

Daily Homework Quiz

1. $1903.74 **2.** $1260.34

3. company A: $P = 15,000(1.2)^t$;

company B: $P = 25,000(1.16)^t$

Lesson Opener

Allow 10 minutes.

1. No; the differences are not the same.

2. Yes; the percent of change from one year to the next is 5%. **3.** about 814,506; about 773,781

4. The percent decrease stayed the same; the percent of change from one year to the next is 20%. **5.** 512; about 410

Practice A

1. B **2.** C **3.** A **4.** $8000 **5.** $5120

6. $3276.80 **7.** ≈$1677.72

8. exponential growth, 1.7, 70% increase

9. exponential decay, 0.2, 80% decrease

10. exponential decay, $\frac{1}{2}$, 50% decrease

11. exponential growth, 2, 100% increase

12. exponential growth, $\frac{7}{6}$, $16\frac{2}{3}$% increase

13. exponential growth, 1.05, 5% increase

14. ≈327,291

15. $P = 2,000,000(0.995)^t$, ≈ $1,902,220$

16. 3000, 2910, ≈2823, ≈2738, ≈2656, ≈2576, ≈2499, ≈2424, ≈2351, ≈2281, ≈2212

17.
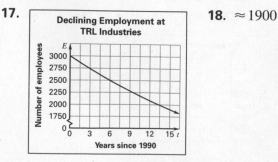

Declining Employment at TRL Industries

Years since 1990

18. ≈1900

Practice B

1. $11,550 **2.** ≈$6848 **3.** ≈$4060.18

4. ≈$1853.60

Lesson 8.6 *continued*

5. exponential growth, 1.25, 25% increase

6. exponential growth, $\frac{6}{5}$, 20% increase

7. exponential decay, 0.94, 6% decrease

8. exponential decay, 0.75, 25% decrease

9. exponential decay, $\frac{4}{5}$, 20% decrease

10. exponential growth, 1.05, 5% increase

11. $\approx 284{,}963$

12. $P = 1{,}400{,}000(0.993)^t$, $\approx \$1{,}305{,}030.08$

13. $\approx 7.0\%$

14.

x	-3	-2	-1	0	1	2
y	128	32	8	2	0.5	0.125

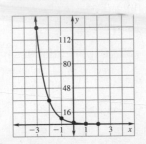

15.

x	-3	-2	-1	0	1	2
y	2.31	1.39	0.83	0.5	0.3	0.18

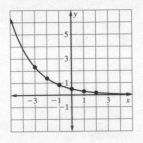

16. 14,000, $\approx 13{,}440$, $\approx 12{,}902$, $\approx 12{,}386$, $\approx 11{,}891$, $\approx 11{,}415$, $\approx 10{,}959$, $\approx 10{,}520$, $\approx 10{,}099$, $\approx 9{,}695$, $\approx 9{,}308$

17.

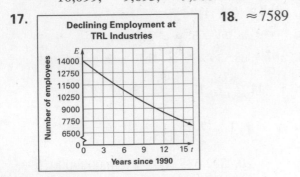

Declining Employment at TRL Industries

Years since 1990

18. ≈ 7589

Practice C

1. $13,770 2. $\approx \$9034.50$ 3. $\approx \$5927.53$

4. $\approx \$3150.13$

5. exponential growth, 10, 900% increase

6. exponential growth, 3, 200% increase

7. exponential decay, 0.75, 25% decrease

8. exponential growth, 1.01, 1% increase

9. exponential decay, $\frac{2}{3}$, $33\frac{1}{3}\%$ decrease

10. exponential growth, $\frac{9}{5}$, 80% increase

11. $\approx 255{,}656$ 12. $S = 8(0.985)^t$, ≈ 5.9 hours

13.

t	1995 ($t=0$)	1996 ($t=1$)	1997 ($t=2$)
E	1285	1255	1227

t	1998 ($t=3$)	1999 ($t=4$)	2000 ($t=5$)
E	1198	1171	1144

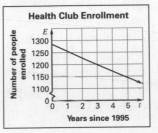

Health Club Enrollment

Years since 1995

14. 14,000, 13,510, $\approx 13{,}037$, $\approx 12{,}581$, $\approx 12{,}141$, $\approx 11{,}716$, $\approx 11{,}306$, $\approx 10{,}910$, $\approx 10{,}528$, $\approx 10{,}160$, ≈ 9804

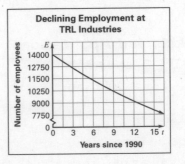

Declining Employment at TRL Industries

Years since 1990

15. ≈ 8204 16. No; You will have $1276.28 at the end of 5 years and $987.56 at the end of 10 years.

Reteaching with Practice

1. $9886 2. $8368 3. $y = 15{,}000(0.90)^t$

Answers

Lesson 8.6 *continued*

4. $7698 **5.**

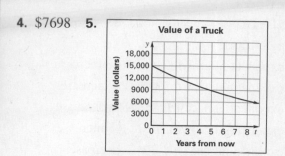

Value of a Truck

Cooperative Learning Activity

1. Answers will vary depending on the student.

2. Answers will vary. **3.** Yes

Real-Life Application

1. a. 55% per year **b.** 1.1 million

2.

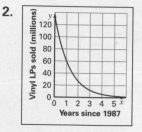

3. Answers may vary. *Sample answer:* Because the cassette tape was the leader for such a short time, people may have been reluctant to upgrade their equipment again from the cassette player to the CD player.

Math and History

1. 1×10^6; 2×10^4 **2.** 5×10^1

3. a. 1×10^{-5} g **b.** 1.9×10^{-8} m

c. 1×10^{-8} cm

Challenge: Skills and Applications

1. $y = 14.7(0.82)^x$ **2.** about 3.5 mi

3. about 3.5 mi **4.** $y = 10(0.5)^{\frac{t}{15}}$

5. about 65 min

Quiz 2

1. 0.000000867 **2.** 7.348×10^7 **3.** 3.6×10^2

4. $1435.63 **5.** 22,514 people

Review and Assessment

Review Games and Activities

1. x^9 **2.** $\frac{1}{3}$ **3.** x^8 **4.** $18x^8$ **5.** 9 **6.** x^7

7. x **8.** $-x^7y^4$ **9.** x^2 **10.** 2 **11.** $\frac{1}{x^9}$

12. $-81x^7y^4$ ABBRA CADAVER

Test A

1. 4^8 **2.** 2^{12} **3.** 42^3 **4.** $4^2x^2y^4$ **5.** b^7; 128

6. a^6; 1 **7.** > **8.** < **9.** $\frac{1}{125}$ **10.** 3 **11.** $\frac{1}{y^3}$

12. $\frac{x^3}{3}$ **13.** A **14.** C **15.** B **16.** 4 **17.** $\frac{8}{27}$

18. $\frac{64}{x^3}$ **19.** $\frac{1}{y}$ **20.** 615 **21.** 0.0114

22. 2×10^{-2} **23.** 1.042×10^3 **24.** 360

25. 20 **26.** $P = 100,000(1.03)^t$

27. $3707.40 **28.** B **29.** C **30.** A

31. exponential growth **32.** exponential decay

Test B

1. 6^{16} **2.** x^{30} **3.** $(-2)^4x^4$ **4.** $3 \cdot 4^2 \cdot x^8$

5. a^6b^6; 64 **6.** $a^{14}b^8$; 256 **7.** < **8.** < **9.** $\frac{1}{16}$

10. 16 **11.** $\frac{y^3}{x^4}$ **12.** $\frac{x^2y^3}{2}$ **13.** C **14.** A

15. B **16.** 4096 **17.** $\frac{49}{64}$ **18.** $\frac{4^4x^4}{3^4y^4}$

19. $\frac{3^3}{2^3x^6y^3}$ **20.** 4326.9 **21.** 0.000071532

22. 3.2×10^{-3} **23.** 3.215625×10^5

24. 0.00042 **25.** 25 **26.** $2519.42

27. $P = 200,000(0.97)^t$ **28.** B **29.** A

30. C **31.** exponential growth

32. exponential decay

Test C

1. x^{22} **2.** $(x - 1)^{24}$ **3.** $(-3)^5x^5y^5$

4. $(-3)^42^3a^{11}b^{15}$ **5.** $-a^{15}b^6$; -64

6. $a^{11}b^{16}$; 65,536 **7.** > **8.** > **9.** undefined

10. 1 **11.** $\frac{8^2}{a^{10}}$ **12.** $\frac{b^{12}}{4^2}$ **13.** B **14.** C

15. A **16.** 1 **17.** $-\frac{64}{125}$ **18.** $\frac{20x^6y^3}{3}$

19. $\frac{2x^3}{3}$ **20.** 892,635,000 **21.** 0.000000016935

22. 1.59×10^{-5} **23.** 1.6826983×10^5

Review and Assessment *continued*

24. 96   **25.** 8   **26.** $P = 250{,}000(1.045)^t$
27. \$8005.02   **28.** B   **29.** C   **30.** A
31. exponential growth   **32.** exponential decay

SAT/ACT Chapter Test

1. B   **2.** C   **3.** D   **4.** D   **5.** A   **6.** A
7. D   **8.** C   **9.** B   **10.** B   **11.** B

Alternative Assessment

1. a. The result is incorrect. When finding a power of a product, find the power of *each* factor and multiply.   **b.** The result is correct. To find a power of a product, find the power of *each* factor and multiply.   **c.** The result is incorrect. When multiplying powers with the same base (such as a), *add* the exponents.   **d.** The result is incorrect. When finding the power of a quotient, find the power of the numerator *and* the power of the denominator.   **2. a.** percents: 3%, −1%, 8%, 4%, 7%, −1%, 5%, 2%, 3%, 4%, 11%, 13%, 9%, 8%, 4%, 4%, 6%, 6%   **b.** Enrollment is increasing overall by about 5% per year.
c. $y = 1414(1.05)^x$; exponential growth; the model represents exponential growth because the enrollment increases about 5% each year.
d. 5543 students   **e.** $y = 1296(1.055)^x$; This model has a similar percent to the model found in part (c), but the y-intercept is different.   **f.** The enrollment will be about 7875 in 2014. Students can use their graphing calculator's table feature to find this solution.   **3.** *Writing* Answers may vary. *Sample answer:* The school district needs this information in order to plan for the years to come. This would include supplies, teachers and staff members, as well as building space.

Project: City Growth

1. Check that the data fit the equations.

2. Check students' predictions and comparisons.

3. Make sure students find linear models for appropriate data sets. Check students' predictions and comparisons.   **4.** Check graphs.

Cumulative Review

1. 27   **2.** −280   **3.** −2   **4.** −18   **5.** $\frac{1}{2}$
6. $\frac{9}{10}$ **7.** 15.2   **8.** −18   **9.** $\frac{17}{5}$ **10.** −42.4

11. 25   **12.** −21   **13.** 2   **14.** −58   **15.** 13
16. 4   **17.** −3   **18.** $\frac{2}{5}$ **19.** $\frac{13}{18}$ **20.** $-\frac{16}{11}$
21. −1   **22.** $-\frac{2}{7}$ **23.** $k = 7, m = 7$
24. $k = \frac{6}{5}, m = \frac{6}{5}$ **25.** $k = 0.2, m = 0.2$
26. $k = -1, m = -1$ **27.** $k = 1, m = 1$
28. $k = 0.23, m = 0.23$
29. x-intercept $= -\frac{7}{6}$; y-intercept $= \frac{7}{2}$
30. x-intercept $= 4$, y-intercept $= 1$
31. x-intercept $= 2$, y-intercept $= -\frac{8}{5}$
32. x-intercept $= 6$, y-intercept $= 12$
33. x-intercept $= 35$, y-intercept $= -0.5833$
34. x-intercept $= \frac{3}{4}$, y-intercept $= \frac{2}{19}$
35. $y = 7x - 6$ **36.** $y = x + \frac{6}{7}$
37. $y = \frac{3}{4}x - 2$ **38.** Mean is 3.5, Median is 2.5, Mode is 2.   **39.** Mean is 19.3, Median is 15, Mode is 6.   **40.** $a = 13, b = 8$
41. $x = 6, y = 7$ **42.** $x = -\frac{21}{10}, y = \frac{19}{5}$
43. $x = -\frac{11}{4}, y = \frac{49}{4}$ **44.** $6561x^8y^8$
45. $(x - 8)^{15}$ **46.** $-3456a^7$
47. $-a^{21}b^{37}$ **48.** $25x^7y^3$ **49.** $-400x^7y^9$
50. $\dfrac{1}{a^6b^{17}}$ **51.** −72   **52.** $\dfrac{8x}{y^6}$ **53.** $\dfrac{5}{r^{14}s^{12}}$
54. 1   **55.** r^7s^2 **56.** $\dfrac{1}{x^5y^{13}}$ **57.** $6x^9y^3$
58. $\dfrac{m^8}{n^9}$ **59.** $\dfrac{2}{5}x$ **60.** $\dfrac{80r^{18}s^{18}}{3}$ **61.** $8x^6y^5$
62. 2.5×10^{-2} **63.** 9.35×10 **64.** 1.5×10
65. 9.9002356×10^7 **66.** 6.3202×10^2
67. 4.631×10^{-4} **68.** \$677.17   **69.** \$1,175.04
70. \$10,451.79   **71.** Exponential Decay. Decay factor is 0.23. 77% decrease.   **72.** Exponential Decay. Decay factor is 0.002. 99.8% decrease.
73. Exponential Growth. Growth factor is $\frac{12}{11}$. 9.09% growth.   **74.** Exponential Decay. Decay factor is 0.56. 44% decrease.   **75.** Exponential Decay. Decay factor is $\frac{8}{15}$. 46.67% decrease.
76. Exponential Growth. Growth is 1.11. 11% growth.